EYEWITNESS TO GEOLOGY

By Steven Lower

In Memory of Kathleen Ann Lower
1947 - 2001

To My Friends and Family
Who Took Me in After My Wife Kathy's Death and
Saved Me from Myself

To Dorothy Parcel, My Friend and Mentor
The Person Who First Encouraged Me to Write

PROLOGUE

During both my education and my career as a professional geologist I was exposed me to a lot of interesting geology. And I saw a lot of geologic disasters, everything from earthquakes to landslides and seacliff retreat.

I started my career as a geologist with the County of San Diego in 1974. In 1979 I went to work for the Kerr-McGee Corporation in Oklahoma City, where I worked for the next 22 years. At Kerr-McGee I did everything from design and construct water supply wells to cleaning up environmental messes.

In retired from my career as a geologist in 2001. During my retirement my interest in geology has continued unabated. I have for many years wanted to write about the after affects of some geologic events in Southern California I had the opportunity to witness for myself, the 1952 Kern County Earthquake, 1971 San Fernando Earthquake, and the 1978 Blue Bird Canyon Landslide in Laguna Beach. And I was able to witness the slow but incessant affects of erosion of the seacliff along a stretch of the San Diego coast.

And when it comes to earthquakes, in recent years Oklahoma has become "earthquake central" in the United States because of earthquakes induced by human endeavors. During my retirement I have continued to monitor earthquakes in the state I have called home since 1979.

TABLE OF CONTENTS

PART 1
LEARNING TO BE A GEOLOGIST

Science, particularly geology, has been an interest of mine for much of my life. After graduating from high school in June 1963 I started college at Fullerton Junior College in my hometown of Fullerton, California the following September. I was interested in science, and of all of my favorite sciences - archaeology, astronomy, anthropology, and geology - the latter was the only field in which I'd be likely to get a job. So I declared that geology would be my major.

Some of happiest days of my life were spent at Fullerton Junior College, first during the middle 1960s and again later in the early 1970s. I thoroughly enjoyed the classes in my major field of geology, as well as other classes I took in the physical, life, environmental and computer sciences.

The junior or community college system provides a two-year, Applied Arts degree. Some two-year programs are designed to provide students with a vocational degree in courses such as nursing or welding or diesel engine mechanic. Other two-year programs are designed to fulfill all of the general education requirements, preparing the student for continuing education at a university. I had intended on going to a university.

All students, regardless of what school they are attending, and what major they are specializing in, are required to take certain "general education" requirements mandated by the state. It is typically better to take those courses close to home. Not only is the cost less at a community college than that in a university, students can typically live at home rather than in a dorm or apartment, further reducing the cost of their education.

During my first semester at Fullerton Junior College I was signed up to take Physical Geology and Physical Geology Lab. I excelled in those classes; I was totally enthused by my first geology classes, and the science of the Earth. I soaked up the geology lectures like a sponge.

What helped fuel my interest in geology was my professor, Walter Reiss. He was an excellent teacher and brought geology to life. Mr. Reiss took me under his wing as my mentor, a situation that was extremely satisfactory to me. He taught me hands-on geology. On field trips he would point out landforms and tell me their relevance to the geology.

At the end of January 1964 I started my second semester at Fullerton Junior College. This semester introduced me to what would be my favorite subject in geology, historical geology. This class mixed my fondness for geology with my other passion, history.

Historical geology is just what the name implies, the study of the history of the Earth. This class involved the study of both rocks and fossils. I was in love with this class! And in addition to the class work there were also three weekend field trips that introduced me to the fun of looking for fossils.

One very exciting thing about the second semester is that I was offered the job of Geology Lab Assistant. This job involved preparing rock and mineral samples and class materials for geology lab classes. I was so excited! I thought of it as a real honor to think that Mr. Reiss thought highly enough of me to offer me such a coveted position. I worked two hours a day, ten hours a week. Not only was I making some money, I was now a member of the Geology Department staff, and I was treated as such. This was my first real job. I kept this job until I went into the Army in March of 1966.

1. The Geology Club: A Gathering of Geeks

Socially I am a nerd, a geek, and very shy. As such I don't socialize much. What was the most life changing experience for me in college, though, was the Geology Club. As a nerd and an introvert, I'm not a joiner. I don't like crowds of people. But because I was geology major, Mr. Reiss encouraged me to attend a meeting of the Geology Club.

Attending that first meeting in the fall of 1963 was a life changing experience for me. While the geology lecture class was full of the "beautiful people," students taking Geology 101 purely to fulfill their lower-division science requirement, the Geology Club was a gathering of students taking geology as either a major or an interest.

The Geology Club should not have been called a club, but rather a "gathering of geeks." And it really was a gathering of geeks and nerds. We were all science nerds. And for the first time in my life I was among my own kind. For the first time in my life I had found people I could talk to, and who wanted to talk to me. And not just at school; we socialized away from school as well. These were people who became good friends, such as Rod Parcel, Bill "Cinnamon Sam" Mallants and Bill Johnson. This was the first time since elementary school that I had good friends, me, Steve Lower, nerd.

Rod's mother, Dorothy, also became a good friend as well as a surrogate mother and another of my mentors. Dorothy invited me into her family and provided me with a second home. There were even a couple of young women, Julie and Terrie, who became friends. Girls who actually wanted to talk to me, the nerdish introvert!

2. Field Trips: Adventures in Geology

Once I joined the Geology Club I became a certifiable "rock hound." And suddenly life became much more adventurous. In addition to class field trips, we members of the Geology Club planned our own field trips. The geology of Southern California is a complex assortment of rock types and geologic settings ranging from recent sedimentary rocks, some bearing fossils, to granites and metamorphic rocks well over a billion years old. It is a geology student's paradise.

The geology field trips were a wonderful new experience for me for two reasons. For one thing, it was hands-on experience: it was great to be able to see and, sometimes, even touch rocks and geologic features we had been studying in class. Secondly, it added adventure and companionship with other nerds who shared an interest common to mine.

After I joined the Geology Club it didn't take long before I started going on rock hunting trips. Rod Parcel had been an avid rock hound for much of his life, an interest instilled in him by his mother Dorothy. In their garage there was no room to park a car; it was full of rock collections. My friends Bill Johnson and Cinnamon Sam Mallants went along on our private rock hunting trips more for the adventure.

At least one weekend a month we members of the Geology Club would arrange for our own field trip. Rod Parcel, Bill Johnson and Cinnamon Sam Mallants and I, sometimes along with others, would go on an expedition searching for rocks and minerals, or sometimes fossils. Rod and I had the oldest cars; his was an early-1950s Chevrolet that, like my 1951 Plymouth, had seen its best days.

If it were just Rod and I, we would look at each car: which one had the best tires, which engine sounded the best, and which had the most gas in the tank. That would be our selection making process. Then we would collect all of the empty soft drink bottles we could find to cash in for the deposit. Coke and Pepsi bottles were glass back then, and had deposits. With gasoline selling for as low as 10 cents a gallon in 1963, four soda bottles would get you a gallon of gasoline. To say that Rod and I we were operating "on the edge of the envelope" would be an understatement.

There were alternatives, though. Bill Johnson had a good job and a nice car, a 1963 Pontiac Grand Prix. Rod and I could usually tap Bill for a ride to our rock-hunting destination as well as for the gas. Rod's mom Dorothy also went along on some of our expeditions, and if she was going it was even better. Dorothy had a 1963 Ford, and was good for the gas as well as food for our rock-hunting trip.

So going on a rock-hunting trip for a day was a big decision making process. But it was always worth it, and always enjoyable if Dorothy went along; I always enjoyed her company.

Steve (left) and Rod Parcel on a
Field Trip to the Desert in 1964

A. <u>The First Geology Club Field Trip</u>

The first, and most memorable, Geology Club field trip was to the Pala Pegmatite District in San Diego County in late September 1963.

This trip held a lot of firsts for me. First it was my introduction to the "geologist's meal in the field." This consisted of summer sausage and summer cheese, both of which needed no refrigeration, and Ritz Crackers. The knife used to slice the summer sausage and cheese was the same used to slice samples of soft rock or split pieces of shale looking for fossils, maybe wiped off on your pant leg. Maybe it wasn't the most healthy or nourishing of foods, but it was very convenient and tasty.

And this was also the first time I had my "rock hammer." This was a hammer made by Estwing that had a point on one end for digging and spitting shale samples looking for fossils and a hammerhead on the other for striking and breaking rocks. Estwing hammers have long been the favorite of geologists because of the quality; hard rocks could be struck without fear that the hammerhead would chip or break. And, of course, I had a canteen for carrying water with me.

That first trip to Pala was my inauguration to being a true geologist.

We were going to Pala in San Diego County to hunt for tourmaline crystals. I had never seen a tourmaline crystal before, so this was another first for me.

Tourmaline is a boron silicate mineral compounded with elements such as aluminum, iron, magnesium, sodium, lithium, and potassium. It is classified as a semi-precious stone, the crystals of which come in a variety of colors. The color of a particular crystal depends on what the boron silicate is compounded with.

Tourmaline Crystals in Quartz Crystals

From the late 1800s through the early 1900s tourmaline was in great demand in China. The Tourmaline Queen Mine was opened in Pala 1903, and was a leading producer of tourmaline through 1914. The Chinese market for tourmaline collapsed in 1911, and by 1914 mining for tourmaline had become uneconomical.

I rode with Rod Parcel and Bill Johnson on the trip from Fullerton to Pala. I felt really excited to be going because for the first time in my life I was a part of something. I was part of a group, even if it was a group of geology geeks. I had found a home.

There were two or three other cars following us with fellow Geology Club enthusiasts, including a couple of girls. On arriving in Pala I discovered that the old mining area was located on lands of the Pala Indian Reservation. Each car had to pay an old Indian two dollars for entrance to the reservation lands.

We followed a dirt road from the highway a mile or so to a shaded area at the end of the road. From here we hiked up the hill quite a ways to where the mines were located. The trail was sided by manzanita bushes that overhung the trail, brushing against you as you walked.

Once at the mining area, which was clear of brush, I learned another first of being a geologist – checking yourself for ticks. These little bloodsucking insects sit on manzanita branches and drop off onto whatever happens to brush against them, man or beast. Everyone, girls included, first checked their clothes for the bugs, then checked around their stomachs. Then we checked our legs, at the tops of our socks. I found one of the insidious insects crawling on my stomach.

Once we were done with our tick check we went into the old tunnel of the Tourmaline Queen mine. To be honest, this is something I would never ever do today, going into the crumbling workings of an old mine. But, in 1963 I was blessed with the stupidity of youth. There is a picture of me squatting in the mine tunnel, with roots hanging from the recently collapsed ceiling.

The tunnel was dug through weathered granite, following a pegmatite dike containing the tourmaline. A dike is a ribbon-like structure that cuts through other rock. The term "pegmatite" refers to the large size of the crystals of rock in the dike.

This was also my introduction to another insect, larger than a tick but just as insidious – the scorpion. Rod had brought the

Geology Department's black light because certain minerals glow in black light. But I found that scorpions also glow in black light, so before you reached for a glowing object you first checked it with a flashlight to make sure it wasn't a scorpion.

Steve in the Tourmaline Queen Mine in 1963

The mine tunnel of the Tourmaline Queen mine was my first encounter with absolute darkness. After a couple of turns in the mine tunnel the darkness is absolute. Rod and Cinnamon Sam turned off their flashlights and wow! Was it ever dark! You literally couldn't see your hands in front of your face. And the girls screamed, which I guessed was the object of turning off the flashlights.

It was cold inside the mine tunnel. It followed one particular pegmatite dike as it dipped down into the earth. The mine walls felt like hard sand, being the weathered granite. In places the sandy surface was replaced by clay, being the weathered feldspar of the pegmatite dike from which tourmaline crystals had been dug out.

Once outside the mine tunnel we started looking through the mine tailings for crystals. It was a clear day, and although the air was cool on that fall day the sun felt hot. As the day started to heat up you could smell the manzanita bushes, a very pleasant odor

Rod and Cinnamon Sam had brought along a couple of sifters, window screen stretched across two-by-fours, which they used to pass the tailings through. It wasn't long before we started to find tourmaline crystals. Most of them were small, but still very pretty. Some were green, some red, and some were what Rod called "watermelon," being a combination of colors. And I found my own very first tourmaline crystals! I was so excited; you would have thought that I had struck gold. Which, in a sense, I had; I was a real geologist, digging for mineral samples!

After searching for tourmaline crystals for a couple of hours we broke for lunch. I must say that my first meal of summer sausage and cheese either straight or on Ritz crackers was very tasty. It would have been nice to have had some wine to wash it down with, but water was all I had. But it made for a good lunch. I felt like I was really "roughing it."

After lunch we went into the Stewart Mine. The entrance to this mine had collapsed, so we had to crawl in under a rock. Again, thanks to the stupidity of youth I was able to do this. Once inside the mine it opened up into a huge, cavernous chamber from which the lithium-rich lepidolite mica had been mined. We found some nice crystals of quartz, and some pieces of pegmatite with the lepidolite mica showing. This was really cool.

Outside the entrance we found something else – and old box full of sticks of dynamite. The sticks were sweating, which Rod said meant they had got hot, but also that it was unstable. We gave the dynamite a wide birth.

Then it was time to leave Pala. We followed the trail back down to the cars, having to check ourselves again for ticks. It had been quite a day, full of adventure and excitement. And I was happy with my finds of little tourmaline and quartz crystals, the start of my rock and mineral collection. I would return to Pala many times over the coming years.

In the early spring of 1964 I camped overnight at Pala with Rod Parcel and Cinnamon Sam Mallants. This was my introduction to another staple of the geologist – beans and franks. We ate "beanie-weenies" cold, straight from little cans. It was the first time I had ever eaten beanie-weenies, and I loved them. I still like beanie-weenies to this day.

B. Memorable Class Field Trips

Because geology was an outdoor activity involving rocks and minerals and fossils as well as geologic features like faults and landslides, there were a lot of class field trips on the weekends. And what wonderful field trips they were! It was very exciting to me to get out and experience geology in its native habitat.

One of the first day trips of my college education was to Long Beach in the fall of 1963. Here was an exposure of sand that had been deposited as part of a beach 20 million years ago. Exposures of this sand were hard to find in the heavily developed area, but were found on one corner of an intersection. Year after year of geology students had stood on this cliff of soft, loose sand, digging out fossil shells that had been deposited on the ancient beach. While the owners of the adjacent home may have looked on with alarm as another group of geology students was actively eroding the lot next to theirs, it was great fun for me. I was actually able to hold 20 million year old shells in my hands!

Field Trip to Long Beach, September 1963

Once I started going on class field trips, especially on the longer weekend trips, I fell in love with the geologic features in Southern California. What especially interested me were the mountains and valleys of the Basin and Range Geologic Province and the Mojave Desert. I had been familiar with the desert of Southern California through camping trips at the Salton Sea south of Palm Springs with my family in the late 1950s and early 1960s. What makes the desert so attractive to a geologist is the lack of vegetation – you can see EVERYTHING. There are no trees and shrubs to mask contacts between different rock formations and structures in the rock, such as faults. Everything is there to see, a fact made plainly apparent by Mr. Reiss during my rides with him on field trips.

Mr. Reiss would point out the window and say, "See that?" "See what?" I would reply. "Where the light colored rocks are displaced downward. That's a fault." And there it was as plain as day. But if Mr. Reiss hadn't pointed it out to me I would have missed it. He taught me to look at things with the geologist's eyes, to recognize geologic features I would have otherwise missed. Suddenly the barren mountains of the Mojave Desert came alive with geologic features. I still to this day, more than fifty years later, look for geologic features in the landscape.

1. The Searles Lake Field Trip; October 1963

My first weekend field trip was in October 1963. This trip was to Searles Valley in the Mojave Desert northeast of Los Angeles. The Searles Lake Gem and Mineral Society hosts a "Gem-O-Rama" on the second weekend of every October, attracting hundreds of rock hounds and mineral collectors from near and far. And beyond the rocks and minerals to be had there is an absolutely fascinating geologic history. It was a "must see" event every year for the Fullerton Junior College Geology Department.

I was really excited to be going on my first overnight fieldtrip. It was the first time I had stayed overnight any place other than at my grandparent's house. Here I was, an 18-year old introvert nerd, actually going away for the weekend! I didn't have a clue as to what I was supposed to be doing, but I was doing it.

Mr. Reiss drove a 1960 Volkswagen beetle. It didn't have many in the way of accouterments: no air conditioning, though to be fair, most cars didn't have air conditioning in 1963. It didn't even have a fuel gauge; if the engine started sputtering he would reach down to turn a lever that opened the reserve fuel tank. The reserve tank held one gallon of gasoline, enough for about thirty miles on the highway. Needless to say, Mr. Reiss kept the main tank topped off.

During all the years I knew Mr. Reiss he drove VW bugs; he loved the cars. He had plenty of money for something fancier, but he loved the look and the simplicity of the VW bugs. To those that knew him, though, they knew that he had another car – a Porsche. But he preferred driving his old VW bugs.

I gained my love for VW bugs from Mr. Reiss. The only new car I have ever owned was the VW bug I bought when I came home from the Army in 1968. I loved that car.

The journey to Searles Lake on a Friday afternoon in October 1963 was a real adventure in itself. Mr. Reiss kept pointing out things to me, features in the landscape, explaining the geologic significance. I learned more during that trip than I ever could have from a book.

We took the highway through Cajon Pass, the dividing point between the San Bernardino Mountains to the east and the San Gabriel Mountains to the west. The San Andreas Fault runs through Cajon Pass between the mountain ranges. Mr. Reiss kept pointing out features in road cuts that showed where the fault zone itself was: a mixed up mess of ground up rock, pulverized by movement along the fault.

This was my introduction to "road cut geology." Due to the courtesy of the highway department, the sides of mountains have been removed to make room for the roadway, exposing the interior of the mountain. Here you could see clearly defined fault zones and the contacts between various layers of bedrock; it was fascinating! Suddenly a barren hillside came alive with geology, as clearly defined as in a photograph. And, in fact, I have taken photographs of many of these road cuts for use in geology classes I taught years later.

Faulted Rocks in a Road Cut

Once through Cajon Pass we were in the Mojave Desert. I had never been on this side of the mountains before, and it was a real experience. Here there was a noticeable change in the air: once out of the coastal basin of Southern California tainted with smog we were in the clear, pristine air of the high desert. After a lifetime of breathing smoggy air, taking in a breath of crisp, clean air was an unforgettable experience.

I learned that the high desert we were in was a "rain shadow desert;" the San Bernardino and San Gabriel Mountains sucked all of the moisture out of the clouds, leaving precious little, if any, for the country on the far side. This was what created the desert – no rainfall.

The highway we were on going over Cajon Pass leads straight east to Las Vegas. We weren't going there, though; that would be a later field trip. So soon after reaching the desert we turned off the highway onto US395, a highway that runs north through the desert.

At one point we stopped at the ruins of an old, abandoned service station and motel called "Esleys." I would swear that this was the same site that Humphrey Bogart stopped at in his 1941 movie "High Sierra." Anyway, here we met Peter Tresselt, the other geology professor at Fullerton Junior College, and his wife Lee, whom I soon learned was nicknamed "Pumpkin." Then we continued north on US395 into the Owens Valley on the east side of the Sierra Nevada Mountains. At the small community of Mojave, more a railroad junction than a town, we turned off US395 and headed east towards Searles Lake. Here my geology education kicked into high gear.

John Searles first noted Searles Valley in 1862. While prospecting for gold, Searles found borax on the surface of the dry desert lake, or playa, in a place he dubbed "borax flats." In 1873, after hearing about the success of borax mining in Death Valley, Searles returned to the valley that would eventually bear his name and filed minerals claims. Searles hauled the mined borax by the famous "twenty mule team" to the railhead at Mojave, from where it was transported to the port at San Pedro. A railroad was later constructed to Searles Valley.

What John Searles had discovered would eventually become known as the world's richest chemical storehouse. In addition to the borax that first attracted him, beneath its surface the valley has half the natural elements known to man in a deposit up to 300 feet in thickness.

Where did all of these minerals come from? To an untrained eye Searles Valley is just another desert valley with a playa, or dry lakebed. But when looked at from a geologist's viewpoint, features stand out that point to a very interesting geologic history. Around the sides of Searles Valley are unmistakable horizontal lines – shorelines of a lake that once filled the valley to a depth of several hundred feet. Look closely at the ancient shorelines and you will see sandy beaches that once graced the shore. Look even closer and you will see sand and gravel bars where small streams had once flowed into the lake from the surrounding mountains. This is just part of what Mr. Reiss showed me as he pointed out the car window.

Suddenly this desolate, hot, dry lake in Searles Valley came alive with a geologic tale waiting to be told.

Searles Valley sits at the western edge of what in known as the Basin and Range Geological Province. This physiographic region includes much of western North America, extending from Oregon south into Mexico. Tensional forces pulling the Earth's crust apart have resulted in the nearly parallel, high mountain ranges separated by deep valleys that characterize the region. The tensional forces spread westward to the Death Valley area two to three million years ago, also creating Panamint Valley and Searles Valley. Death Valley is the deepest of these fault-block valleys, today lying 280 feet below seal level.

The Sierra Nevada Mountains lie about 40 miles west of Searles Valley. Granite makes up most of the bedrock on the eastern side of the mountain range, along with some metamorphic and volcanic rocks. Minerals in these rocks contain virtually all of the elements known to man, including some radioactive elements such as uranium.

During the last Ice Age the environment in California was much wetter and colder than today. Rivers of ice called glaciers filled many of the valleys of the eastern Sierra Nevada Mountains. These glaciers moved slowly down slope to the east off the mountains, grinding the granitic bedrock over which they moved into a fine powder called "glacial flour." During "interglacials," short periods of warmer temperatures when some of the glacial ice melted, streams carrying the glacial flour flowed out of the mountains and into the Owens River east of the mountains.

The Owens River drains the eastern Sierra Nevada Mountains, carrying with it the glacial flour. The Owens River flows south through Owens Valley, then east through a series of five interconnected desert basins. Owens Lake is the first of these basins. During periods of high glacial melt, Owens Lake would overflow into the next basin, China Lake, and from there into the third basin, Searles Lake.

During the last 2½ million years Searles Lake was normally the end of the Owens River, but during periods of heavy glacial melting it overflowed into Panamint Valley to the east, which in turn overflowed into Death Valley. This was the end of the chain of five lakes, creating Lake Manly.

With no outlet other than overflowing into Panamint Valley, water sitting in Searles Valley slowly evaporated over tens of thousands of years. Every time Searles Lake would shrink in size it would leave behind minerals from the settled glacial flour in an ever-increasing concentration. As the lake continued to shrink in size the remaining water would eventually become brackish, heavily laden with dissolved minerals. If the lake dried up completely it would leave behind thick layers of mineral-laden salt and mud.

Studies of Searles Lake have shown that a total of thirty lake level cycles had occurred during the last 150,000 years alone. Those lake levels are the horizontal lines seen on the mountainsides.

Each of these cycles took thousands of years. The result is an accumulation of brine-laden sediment up to three hundred feet in thickness. The last cycle occurred 10,500 years ago. The ice age ended and the climate turned hot and dry. Searles Lake dried up as the area turned to desert.

The company that owns Searles Lake, Searles Valley Minerals, pumps the brine to the surface where the valuable minerals are extracted. The leftover water is pumped back into the lake sediment where is will dissolve more minerals, in a process called solution mining.

One noticeable thing when we drove into Searles Valley was the smell. It smelled like evaporated sea water, that "briny" smell. And that was what it was – brine, heavily concentrated saltwater. And there was also the smell of the salt. You wouldn't think that salt has a smell, but when you have millions of tons of the stuff it has a smell.

I learned to associate these smells with Searles Lake. In subsequent years when going to Searles Lake for the annual Gem-O-

Rama I would get excited to smell the odor of the lake, something I looked forward to.

My friends Rod Parcel and Bill Johnson had ridden to Searles Lake with Rod's mother Dorothy. This being my first ever weekend trip away from home I didn't have a clue as to what I was supposed to be doing, so I glued myself to them, and Dorothy took me under her wing and looked after me as the naïve boy I was, camping in the desert without a clue as to what I was doing.

While in the town of Trona at Searles Lake that evening I had stocked up on some food while Mr. Reiss filled his gas tank. If you could call it food – a big bottle of Coke and packages of Hostess Cupcakes. Like I said, I didn't have a clue, and knew nothing about fending for myself. Dorothy took pity on me and mad sure I was fed.

Friday night was a unique experience for me. Hundreds of people filled the campground set aside for people attending the Gem-O-Rama. It was a primitive site in the desert, no water and no facilities other than some outhouses. But it was very much a party atmosphere. People sat around drinking beer and talking in the warm October evening.

One thing I noticed about the desert was that once the sun started going down the air cooled quickly. I had been told to bring a light jacket, and I'm glad I did. Even though the ground still felt warm, the air got rather cold that night.

There were a lot of geology students from Fullerton Junior College, as well as graduates. Other colleges were well represented as well. For many it was a homecoming of sorts, graduates returning to Searles Lake to meet up with old friends as well as past professors. In later years I would be doing the very same thing. Then there were the people who were just interested in collecting minerals. I was in good company, being among nerdish geology enthusiasts.

Yet as an introverted nerd I was too shy to talk to any of them. I've never been good at striking up conversations with strangers; I'm just too shy

Beyond being alone among hundreds of people I didn't have a clue as to what I was supposed to be doing while at Searles Lake. I was completely lost. My parents had bought me a sleeping bag to use on the trip, but other than that I had no camping gear. I was totally unprepared. But Dorothy steered me in the right direction and fed me.

During that Friday night I was amazed at the sky: filled completely with stars. The Milky Way was pointed out to me; I think it was the first time I had seen the Milky Way.

I stayed close to Dorothy and her family that night. Saturday morning broke with a hot sun in a clear sky. Mr. Reiss cooked us breakfast – cheese omelets, the first ones I'd ever had. It was a real treat for this kid on his first overnight trip from home.

After breakfast it was out to the salt-crusted lakebed to look for crystals. During the annual Gem-O-Rama, rock hounds are given access to part of the lake surface to collect mineral samples. Some of the supersaturated brine is pumped to the surface, where the minerals are allowed to crystallize, which takes only minutes.

Crowd Watching A Brine Pumped from Searles Lake

It was very much a party atmosphere as people excitedly sifted through the crystallizing brine. You saw people of all ages, from young children to old folks excitedly searching for crystal treasures.

Crystals of the mineral Halite, or Sodium Chloride, also known as common table salt, were by far the most common. Crystals of the Potassium-Sodium Sulfate mineral Hanksite were the next most common finds, although much more rare than the salt crystals. The most coveted crystals of all were those of the Sodium Carbonate mineral Trona. While I did find some Hanksite crystals, on all of the Gem-O-Ramas trips I went on over the years I never did find a Trona crystal.

Hunting Minerals at Searles Lake

I had a lot of fun just watching the other people as they hunted crystals. One person had on a wide brim hat on with a little fan below the front that blew air on their face. Since the air was very still, with nary a breeze, that seemed like a good idea.

The brine on the surface saturated your clothing real fast. I had also been told to bring along an extra set of clothes, and again I was glad I did. My socks and legs of my jeans were stiff as boards from the salt encrusting them. But all in all it was a lot of fun.

By noon, it was getting too hot to continue the hunt for mineral crystals. Even in October the sun can be racing past 90 degrees at noon, much too hot to be out on the white lake surface. So after lunch it was off to hunt for more geology.

Steve Collecting Minerals at Searles Lake
October 1963

I again rode with Mr. Reiss as we drove into Panamint Valley, the long, narrow valley located between Searles Valley to the west and Death Valley to the east. A major cultural feature in Panamint Valley are the ruins of a once thriving community, Ballarat. Today it is a "ghost town," but it once had a population of up to 500 people.

During the late 1800s there was a lot of gold mining activity in the Panamint Mountains bordering the east side of the valley. Ballarat was founded in 1896 as a supply point for the miners. During its heyday from 1897 to 1905 the town had seven saloons, three hotels, a Wells Fargo station, a post office, and a jail. It was a place where miners and prospectors could go to relax or to buy provisions. After 1905 the mines began to play out; the last mine quit operations in 1917.

Later Saturday afternoon we drove back into Searles Valley. After a visit to the Searles Valley Gem and Mineral Society's museum in Trona it was back to the campground for the night.

Saturday night I stayed close to Dorothy Parcel and her family again, and really enjoyed all the stories about geologic adventures that were told that night. Sunday morning we packed up and started on the long journey back to Fullerton. Even though I was a totally lost and clueless nerd, I would have to think of this first overnight camping trip as a success.

The trip to Searles Lake was the highlight of my first semester at Fullerton Junior College. I'll confess that I wasn't the best of students during my first semester in college. I got an "A" in geology, but "Bs" and "Cs" in everything else. Not a stellar start to my college career. I just didn't do as well as I would like to have done. I had the smarts; I just didn't have the interest. My biggest problem was that I didn't apply myself as well as I should have in classes I didn't like. Later on, when I had more control over what classes I took, my grades were much better. I promised myself that I would do better in my second semester at Fullerton Junior College.

2. Field Trip To The Marble Mountains

As with my first semester at Fullerton Junior College, there were field trips with my Geology Club friends. My most memorable fieldtrips, however, were with my historical geology class. Of course, my Geology Club friends also went on these trips.

The first class field trip of my second semester at Fullerton Junior College was to the Marble Mountains near the small desert community of Amboy. There we visited one of the most famous fossil hunting areas of Southern California, the trilobite fossil locality. While in the area we also visited the Amboy volcano crater. The crater is located a couple of miles from the town of Amboy on historic Route 66.

This field trip didn't have the party atmosphere, as did Searles Lake; it was all about learning geology.

We left Fullerton Junior College Saturday morning and drove south towards Palm Springs. The fall of 1963 was well before construction of the Interstate 15 connecting Riverside to the north and Palm Springs to the south through Banner Canyon. In 1963 the route was still a two-lane highway.

After Banner Canyon we headed east into the southern Mojave Desert towards Amboy. After crossing a mountain range a long, broad desert valley lay in front of us. In the distance we could see salt flats and, north of that, the unmistakable shape of a volcano surrounded by the dark rocks of the lava flows.

Amboy Crater, described as a "cinder cone" type of volcano, is about 80,000 years old. Being 250 high above the desert floor and surrounded by a lava flow covering 27 square miles, the crater is the most distinctive feature in the area.

In the fall of 1963, US Route 66 was still the only highway running southeast through the Southern California desert. The scenic Amboy Crater was a popular sight and stop for travelers on U.S. Route 66 before the opening of Interstate 40 in 1973. The crater and its surroundings have been used as settings for several motion pictures as well as commercials over the years.

Amboy Crater and Lava Field

US Route 66, which runs north to south through the valley before turning east towards the Arizona border, follows the route of the San Fe Railroad through the southern Mojave Desert. During the railroad building years in the 1880s a new town was built about every twenty miles along the route. Towns such as Needles and Baker and Amboy and Barstow were all towns built by the railroad. When Interstate 40 was completed, replacing the southern US Route 66 as the only highway, the towns that were bypassed went into decline.

We stopped where the highway cut across the lava field, and then hiked to the crater. The lava that erupted from Amboy Crater was a very thick, viscous variety that results in a lava field with a very hummocky surface. A lucky find in the lava field is a "lava bomb," a blob of molten lava that erupted out of the volcano and flew through the sky. It cools and hardens quickly as it flies through the air and takes on an aerodynamic shape.

The last eruption of Amboy Crater was especially explosive, throwing out lots of lava bombs, and blowing out the south side of the crater in the process.. Inside the volcano can be seen the frozen remains of what had been the lava pool.

Inside Amboy Crater Showing the Lava Pool

Saturday evening we drove into the town of Amboy for gas and a bite to eat. Amboy was founded by in 1883 as part of a series of railroad stations that were to be constructed across the Mojave Desert. In 1926 Amboy became a boomtown of sorts following the opening of U.S. Route 66.

Roy's Motel and Café opened in Amboy in 1938 and has been a fixture in the tiny town ever since. Roy's was the only gasoline, food and lodging stop for miles around that part of the Mojave Desert. It was a long way between towns when driving across the desert, and you always needed to make sure you had the necessary gas, food and water to make it to the next town.

We ate at Roy's that Saturday night, nothing special, just food. Then we camped out near the crater. I was a little better prepared this time; my parents had bought me a "pup tent" for my birthday, a simple little canvas tent. It was a good size for one person, and for me it was like being in my own little clubhouse. I had bought some goodies at Roy's, candy bars and Hostess cupcakes and a coke; I still didn't have feeding myself down as well as I should have. Besides, I have always had to have my secret stash of goodies in my "clubhouse," and I am still that way today, 50 years later.

Early the next day we went to the Marble Mountains fossil beds. The Marble Mountains have been a favorite fossil-hunting site for generations of geology students as well as fossil aficionados. Here can be found abundant and well-preserved fossil trilobites. These creatures date to the early Cambrian geologic age, or roughly 518 million years ago, and are some the earliest animals with preserved hard parts. Their name comes from the fact that their bodies were comprised of three parts, or lobes, of an exoskeleton: a head, a thorax or middle part, and an abdomen or tail end part.

Trilobites are early relatives of later arthropods such as crabs as well as insects such as scorpions.. The trilobites were thought to have been predators and scavengers, scurrying across the bottom of shallow seas feeding on plankton and whatever else they could find.

The fragments of the hard, chitinous exoskeleton are what are preserved as fossils. The most common finds are the remains of the head coverings, or carapaces. Lucky is the fossil hunter who finds a fossil of the entire animal.

The trilobite fossils are found in the Latham Shale of early Cambrian age. The Latham shale was formed from the muddy bottom of a shallow sea not far from the coast of what would eventually become North America. In early Cambrian time this area was at the equator. The trilobites lived in a warm, shallow tropical sea that provided the perfect environment for the proliferation of early life.

Well Preserved Trilobite Fossil

The shale occurs in thin, hard layers that can be easily split apart along natural bedding planes. Whenever splitting a piece of shale one always hopes to find a magnificent trilobite fossil inside.

Trilobites were a very successful species, existing for nearly 270 million years. They died out during the great Permian Extinction Event about 250 million years ago, before even the dinosaurs roamed the land.

Being in the Mojave Desert, the best times to visit the Marble Mountains is anytime other than summer. Even on an early spring day the temperature can easily reach into the 90s in the sun. And the relentless sun does beat down on you, so a good hat is an absolute necessity. And the air is dry, which together with the heat means you need to take along lots of water. Beer would be even better, but it is a long walk from the parking area to the fossil location to be carrying an ice chest. Even so, the absolute thrill of finding trilobite fossils, even just the heads, overcomes any and all discomfort. And I was thrilled to find trilobite fossils.

Fourteen years later, when I was teaching at San Diego Mesa Community College, I took my own geology students to the Marble Mountains. And they, like me, thoroughly enjoyed the experience of collecting the fossils of creatures that lived 518 million years before.

***Steve's Students Hunting Fossils at the
Marble Mountains in 1978***

3. The Great Las Vegas to Death Valley Field Trip

This was one of the most memorable trips of my life. Even though I have followed this same route a few more times since, this first trip was unquestionably the most memorable.

On this trip I rode with a fellow student named Rene Pence. He had a fairly recent and nice car. Being about 40 years old he was an "old man;" I had just recently turned 19.

We left Fullerton Junior College Friday evening and headed to Las Vegas. I was really excited; I had never been to Las Vegas before. I was 19 years old and going to Las Vegas!

The trip seemed long across the Mojave Desert. This was 1964, long before Interstate 15 had been built; the highway was still two-lane with a 55 miles per hour speed limit. Until you reached the Nevada border, that it. As I was to learn there was no speed limit in Nevada.

It was well after dark on a Friday evening as we were driving over a mountain range. Ahead I could start to see a glow in the night sky. Once we crested the mountain range the valley that lay ahead was awash with light, just a great big ball of light. It seemed a long way still to the outer reaches of Las Vegas when you could start to make out individual lights. All I could say was WOW!

I would suggest to anyone driving to Las Vegas to plan to arrive after dark so you can experience the magic of the great light show.

I was just in absolute wonder as we arrived in Las Vegas. And mind you, this was in 1964, when what was then the city limits is now close to the middle of the city, it has grown that much.

We had arranged to meet at the Desert Inn Hotel and Casino. This was my first ever visit to a casino, and I was in awe! All of the lights, all of the slot machines, all of the noise. This was when slot machines were really "one armed bandits;" you pulled the lever to spin the wheels. This was also when the nickel was still the coin of the realm when it came to slot machines. Unlike the slot machines of today, when you heard the clatter of coins falling into the hopper it wasn't a sound effect – real coins were falling into the hopper. And you could see people walking around with small buckets filled with coins.

I had just turned 19, but I looked much older, and had no trouble getting into the casino.

I stayed glued to Rene, the old guy who had a clue as to what was going on. This was thirty years before cell phones, so if you got lost you were REALLY lost. We met up with Walter Reiss and the other students in the casino. And the first stop of this field trip was in the men's room of all places (there were no girls in this class). The walls between the stalls were the attraction – they were full of fossils! We were told that these walls were made of a limestone dating to the Cretaceous era, roughly 80 million years old. The fossils were of Turritella shells, a type of gastropod, or snail, that lived on the bottom of shallow seas.

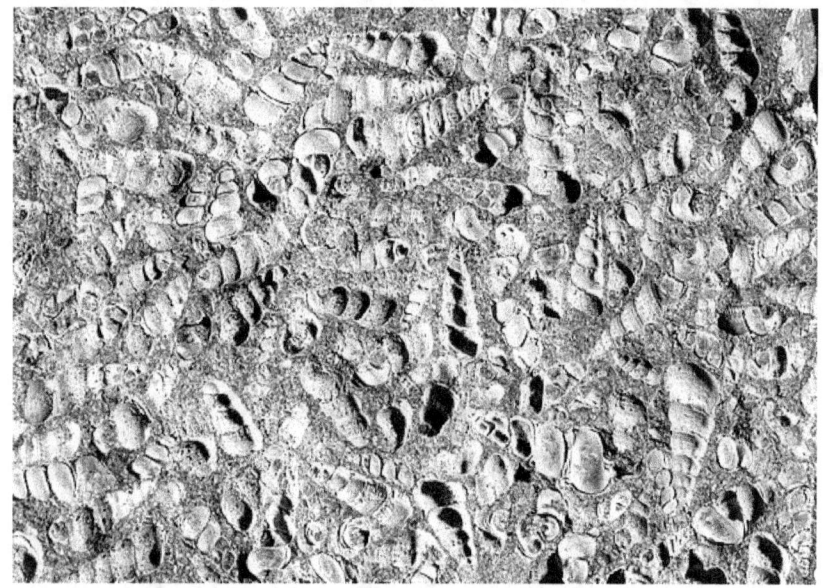

Fossil Turritella Shells in Limestone Wall

After this tour of the men's room we had dinner at the hotel's smorgasbord, a cafeteria that had any food you would like to have. Again I was in awe – I had never seen a smorgasbord before. After dinner we gambled in the casino for a while, dropping nickels into the slot machines and pulling the lever to spin the wheels. Then it was out to the desert to camp for the night. I had my sleeping bed and pup tent again.

The next morning we drove back into town to have breakfast at the breakfast bar at one of the casinos. And again I was in awe – so much food at such a low price! I had never known that places like these even existed.

After breakfast we headed north out of Las Vegas north over the Spring Mountains. We stopped at a road cut to sample rocks that we were told belonged to the Monte Cristo Limestone. These carbonate rocks were deposited during Mississippian time some 300 million years ago. At that time this part of Nevada was near the coast of North America, with this particular area part of the continental shelf under warm, shallow seas.

From the Spring Mountains we continued north to Amargosa Valley. Here we turned and headed southwest towards Death Valley. We stopped at the old mining town of Ryan, then at a spectacular overlook of Death Valley at Zabriskie Point. The viewpoint at Zabriskie Point offers a wonderful view of Death Valley. Hear we learned how Death Valley formed just a few million years ago.

Steve at the Zabriskie Point Overlook

After stopping in a store in Stovepipe Wells in Death Valley we camped out for the night. After the smorgasbord at the Desert Inn this was really roughing it; more beanie-weenies. But I survived. As at the Searles Lake campground a few months before, I was awestruck by all of the stars in the sky.

An interesting memory about Stovepipe Wells is buying an ice cream bar at the store. It was vanilla ice cream covered in chocolate. Even in late April the temperature was high in Death Valley. Once I walked out of the air-conditioned store the ice cream started melting so fast that the chocolate coating just slid right off the bar and all over my hand. I learned a valuable lesson that day – you can't eat chocolate covered ice cream bars in Death Valley.

The next day, Sunday, we drove west through Panamint Valley, through Searles Valley, then there was the long drive home. Thus ended one of the most exciting trips of my life. And my first of many visits to Las Vegas.

These field trips were the highlights of my second semester at Fullerton Junior College. A couple of years later, in 1966, the Vietnam War interrupted my college education.

After my service in Southeast Asia I went back to school at Fullerton Junior College in the spring semester of 1971. I wanted to see if I still had what it took to go to school, and if a college education was in my future. If so, I would need to get the credits necessary to transfer to a four-year college. The government would pay me to go to school on the GI Bill, the one good thing I got out of the Army.

Also at this time, in 1971 and 1972, I was given back my old job as part-time technician in the Fullerton Junior College Geology Department's lab, the job I had left in 1966. As the lab assistant I went on most of the geology class field trips.

I graduated from Fullerton Junior College in 1972 with a two-year Associate of Arts degree. Throughout my continuing education there would be lots of fieldtrips. In the late 1970s I took my own students on field trips when I was teaching geology at San Diego Mesa Community College. But those first Geology Club and class field trips at Fullerton Junior College in the middle 1960s will always remain as the most memorable.

At the recommendation of a good friend I transferred to Long Beach State University in September 1972 as a junior to finish my degree in geology. I spent one school year at Long Beach State, and then decided to transfer to San Diego State University. I started at San Diego State as a senior in September 1973.

At the suggestion of my undergraduate advisor I specialized in the hydrogeologic side of geology, the study of geology and groundwater; I began studying streams and springs in the Laguna Mountains east of San Diego. I started collecting data in the fall of 1973, on which I based my undergraduate thesis. I finished my bachelors degree in hydrogeology in June 1974, and started on my master's degree the following September.

When I started my graduate program in September of 1974 the subject of my graduate thesis was already decided - I continued and expanded on the groundwater work I had done my undergraduate thesis on. At least once every month for two years I went into the mountains to collect data on streams and springs.. By the time I finished the fieldwork for my thesis I had more than three years of data to work with. My resulting thesis had the distinction of being one of the longest ever produced by the San Diego State geology department.

In the days before personal computers and word processors and spreadsheets and MS Windows, writing a report of more than 200 pages, filled with pictures and maps and graphs and tables, was a very long and tedious process. I bought an electric typewriter and built a light box for tracing maps. A single mistake in the report text could (and did) result in the retyping of a large part of the report.

A professional typist prepared the final text. For a thesis there is a long process of getting it finalized and approved. With the final approval of the thesis in November 1977 I had completed everything for my master's degree. Although I wouldn't go through a formal graduation ceremony until the end of the school year in June 1978, my diploma is dated December 1977.

Because of the groundwater work, in September 1974 I got a job with the County of San Diego's Integrated Planning Office (IPO). The IPO was working on the implementation of the California Environmental Quality Act (CEQA) in San Diego County. My assignment was to study the geology and groundwater availability in sub-regional community planning areas and make recommendations on permissible population densities given the groundwater resources available.

I spent a lot of time in the rural areas of San Diego County studying the geology and measuring water levels in water-supply wells. I eventually wrote reports on groundwater availability in the Jamul, Ramona, and Julian Subregional Growth Management Areas. I also made presentations on groundwater to community groups, which was a lot of fun. This was the start of my career as a hydrogeologist, which would eventually lead me to Kerr-McGee Corporation in Oklahoma City in 1979. I retired from Kerr-McGee in 2001. Even though I may no longer be working in the field, geology has continued to be an interest to me during my retirement.

PART 2
EARTHQUAKES

An earthquake, also known as a quake, tremor or temblor, is the noticeable shaking resulting from seismic waves caused by the sudden release of energy when the Earth's crust ruptures. The seismicity or seismic activity of an area refers to the frequency, type and size of earthquakes experienced over a period of time.

A note about the magnitude and energy of earthquakes. The Richter magnitude scale assigns a magnitude number to quantify the energy released by an earthquake. The Richter scale, developed in the 1930s, is a base-10 logarithmic scale, which defines magnitude as the logarithm of the ratio of the amplitude of the seismic waves.

As measured with a seismometer, an earthquake that registers 5.0 on the Richter scale has a shaking amplitude 10 times greater than that of an earthquake that registered 4.0.

As a point of reference, an 8.6 magnitude earthquake releases the same amount of energy as 10,000 atomic bombs like those used against Japan in World War II.

At the Earth's surface, earthquakes manifest themselves by shaking and sometimes displacement of the ground. Earthquakes can also trigger landslides, and are commonly associated with volcanic activity. When the epicenter of a large earthquake is located offshore, the seabed may be displaced sufficiently to cause a tsunami, as did the 2004 Indian Ocean earthquake and the 2011 earthquake off the coast of Japan.

Earthquakes cause damage two basic ways. One way is shaking, caused by the rapidly moving peaks and valleys of rolling earthquake waves moving through the ground. Shaking can also cause the soil a structure is built on to sink due to compaction.

The second way earthquakes cause damage is through rupture of the ground along fault zones. This rupture may be through either vertical or horizontal movement of the ground or a combination of the two.

You can actually hear an earthquake coming – it sounds like the low frequency rumbling of an approaching freight train.

An earthquake's point of initial rupture is called its focus or hypocenter. The epicenter is the point on the ground level directly above the hypocenter. Earthquakes are measured using observations from seismometers, measured mostly on the Richter magnitude scale. Earthquakes of magnitude lower than 3 are generally imperceptible to most people, while earthquakes of magnitude 7 and above can cause serious damage over large areas. The largest earthquakes in historic times have been of magnitude slightly over 9, although there is no limit to the possible magnitude.

Earthquakes are typically thought of as natural or tectonic events occurring along fractures in the Earth's crust, called faults, but some can be triggered by human activity. The injection of fluids into fault lines can cause slippage, triggering earthquakes. Nuclear tests and mine blasts can cause localized seismic activity.

Tectonic earthquakes occur when two adjacent blocks of the Earth's crust slide by one another. If the crustal blocks slide by one another easily in a constant motion there may be little if any seismic activity related to the movement.

The San Andreas Fault in California marks the junction between the Pacific Ocean crustal plate to the west and the North American plate to the east. These two tectonic plates are slowly moving past one another as the North American plate pushes westward. There are places along the San Andreas Fault in the San Francisco area where the two plates smoothly "creep" by one another at a rate of a fraction of an inch a year, slowly offsetting everything from streets to tennis courts that are built across the fault line.

There are other areas along the San Andreas Fault where the two plates are "locked" in place. Because the Pacific and North American plates are still trying to move past one another at the locked section, stress builds until the rocks of the adjoining plates along the fault line "break," suddenly lurching past one another in what may be inches or even several feet at one time and causing an earthquake This process of gradual build-up of strain and stress punctuated by occasional sudden earthquake failure is referred to as "elastic rebound."

***Trace of the San Andreas Fault through the
Carrizo Plain of California***

Faults are classified as three different types: normal, reverse and strike-slip. Normal and reverse faults are examples dip-slip faults where movement is in the vertical direction. Normal faults occur mainly in areas where the crust is being extended or pulled apart; these kinds of faults often result in valleys down-dropped between mountain ranges.

Reverse faults occur in areas where the crust is being compressed. Thrust faults are an extreme reverse fault where the crust on one side of the fault is being pushed up and over the other side. Mount Everest in the Himalayan Mountains is an example of a "megathrust;" the Indian Subcontinent is being thrust over the Asian Continent, pushing the mountains up in the process.

Strike-slip faults occur where the crust on both sides of the fault slip horizontally past one another; the San Andreas Fault is an example of the strike-slip fault. Many earthquakes are caused by movement on faults that have components of both dip-slip and strike-slip, resulting in what is known as oblique slip.

Reverse faults, particularly those along converging plate boundaries where two crustal plates are pushing against one another, are associated with the most powerful earthquakes, called megathrust earthquakes, including almost all of those of magnitude 8 or more. The 2004 Indian Ocean earthquake was a megathrust earthquake. Strike-slip faults can produce major earthquakes up to about magnitude 8. Earthquakes associated with normal faults are generally less than magnitude 7. This hierarchy of earthquake intensity is due to the causative stress levels in the three fault types. Thrust faults are generated by the highest, strike slip by intermediate, and normal faults by the lowest stress levels.

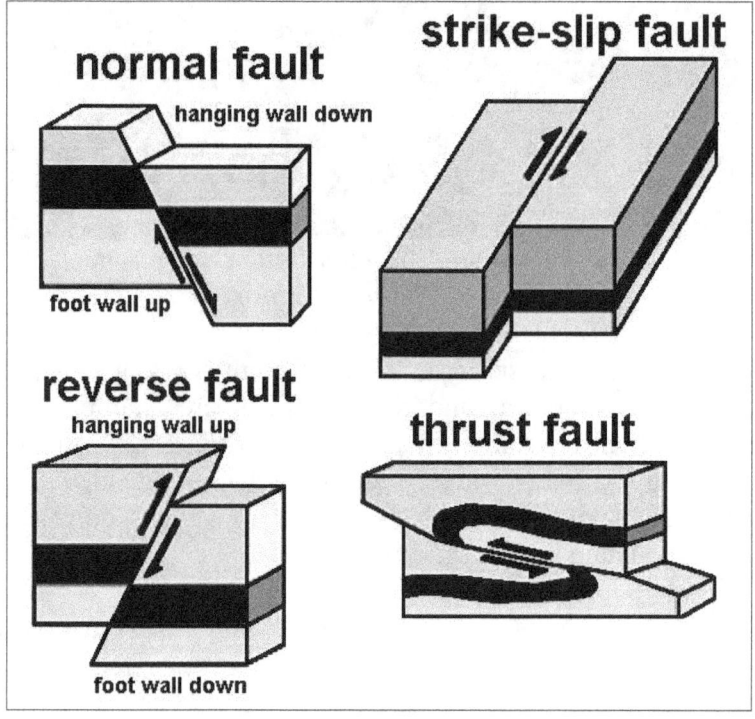

Types of Faults

Most earthquakes have "Intraplate" origins, occurring along boundaries between tectonic plates, such as the San Andreas Fault. Intraplate earthquakes, those occurring within an individual tectonic plate, are typically rare and of small magnitude. This has changed in recent years due to earthquakes caused by humans; earthquakes occurring in Oklahoma, in the middle of the North American plate, are a textbook example of induced seismicity.

Example of a Reverse Fault

Earthquakes can occur in a series of events called an "earthquake swarm." Over a period of days or weeks there will be a series of earthquakes involving a single fault line. After all is said and done the biggest of the earthquakes will be defined as the "mainshock," while those before will be defined as "foreshocks." Likewise, those that came after the main shock will be defined as "aftershocks."

1. The Kern County Earthquake

I was born in Pasadena, California in 1945. I lived in California until my wife and I moved to Oklahoma in 1979. I experienced a lot of earthquakes during those 34 years I lived in California, mostly small tremors, but a few fairly big.

The first earthquake I remember as a child was in July 1952 when I was seven years old. I clearly remember being in our Pasadena house when everything started shaking very noticeably, the ground rolling and the living room chandelier swinging. My father took me out into the front yard where I remember seeing fire and sparks coming from the power poles and lines.

I remember seeing damage on the television (yes, we had television in 1952, albeit an 8-inch black and white screen), with scenes of rows of homes knocked down by the earthquake.

Little Boy Steve Lower

Twenty years later, while in college studying to be a geologist, I learned that this was the Kern County Earthquake of July 21, 1952. It was fairly large earthquake, with a magnitude of 7.3 that killed 12 people and injured 18, and caused an estimated $60 million (in 1952 dollars) in property damage. It is also called the Tehachapi

Earthquake because the town of Tehachapi, located in the mountains north of Los Angeles, suffered the greatest damage and loss of life. At the time it was the strongest earthquake to occur in California since the great San Francisco earthquake of 1906.

I also learned that ground movement during an earthquake can cause power lines to swing together and make contact, shorting out transformers, causing them to literally explode in flames. That was the source of the sparks and flames I had seen on the power poles and lines in 1952.

The epicenter of the Kern County earthquake was in the southern end of the San Joaquin Valley of central Southern California. The hypocenter of the earthquake was at a depth of 9.9 miles at on the White Wolf Fault near the community of Wheeler Ridge. The White Wolf fault has left-lateral movement, meaning the block on the opposite side of the fault moves to the left in the horizontal plane. In this case, though, there was also a vertical element to the rupture along the fault, making it an oblique-slip earthquake.

The July 21 mainshock had a significant aftershock sequence that persisted through September 1952. Through late September seismometers had recorded 188 aftershocks higher than magnitude 4.0. Six of those on the day of the mainshock were magnitude 5 and above, but some of the smaller ones were only felt, and didn't cause any damage. The strongest aftershock came on August 22 as a magnitude 5.8 event that resulted in the deaths of two people and caused an additional $10 million (1952 dollars) in property damage.

Though damage from the Kern County Earthquake and its aftershocks was spread throughout a large area of central and southern California, most was concentrated in the town of Tehachapi. A total of 700 families were affected in Tehachapi alone, where most of the town's buildings sustained damage. In Bakersfield, near the southern end of the San Joaquin Valley, windows were broken and plaster fell from walls.

Due to its distance from Tehachapi, most of the greater Los Angeles Area was escaped serious damage. Power disruptions affected Van Nuys, Los Angeles and Pasadena, due partly to the transformer failures I had seen.

Rows of cotton in San Joaquin Valley were offset along the rupture of the fault line. In the same area an east-west road was dislocated five feet by movement on the fault. Near the mouth of Comanche Creek a shallow-sloped fault scarp was raised with a maximum vertical displacement of three feet, reflecting the oblique-slip nature of movement along the fault.

Damage Caused by the July 1952 Earthquake

Railroad Tracks Displaced by the July 1952 Earthquake

The American Red Cross declared the Kern County Earthquake a major disaster, but getting relief into the area was stalled because of landslides blocking the highway running between Los Angeles and Kern counties. Two tunnels used by the Southern Pacific Railroad and the Santa Fe Railroad collapsed between the towns of Tehachapi and Marcel, halting rail transport of relief supplies.

2. The San Fernando Earthquake

Damage to Part of Veterans Hospital in Sylmar, California
February 9, 1971

About 6am on the morning of February 9, 1971, I was jostled awake and nearly thrown from my bed by the severe shaking of my bedroom. I immediately knew it was an earthquake, a big one. I got up and turned on the television and watched as news reports started coming in about a major earthquake in the San Fernando Valley, located north of Los Angeles in southern California, about forty miles north of my Anaheim apartment.

The first news reports focused on the Van Norman dam, which had suffered a partial collapse. Thousands of people living in the valley below the dam were being evacuated as a precaution in case the dam was overtopped by water. As time went by reports started coming in about severe damage to the Veterans Administration hospital in the community of Sylmar in the San Fernando Valley.

This was the San Fernando Earthquake, also known as the Sylmar Earthquake, which was centered at the north end of the San Fernando Valley in the foothills of the San Gabriel Mountains.

In early 1971 I was back in college as a geology major at Fullerton Junior College in Fullerton, California. I had first started at Fullerton Junior College in September 1963 but had my education interrupted by the Vietnam War. About a month after the February 9,

1971 earthquake, the Geology Department organized a field trip to the San Fernando Valley to see the damage and it's geologic impacts first hand.

Collapsed Concrete Wall of the Van Norman Dam

Foothills of the San Gabriel Mountains with Sylmar in the Foreground

The damage from the earthquake was shocking and horrifying in its scale and severity. A total of 64 people died during the earthquake and its more severe aftershocks. In total the earthquake caused more than $500 million in damage (1971 dollars). The damage from the 1952 Kern County Earthquake, while a much larger seismic event than the San Fernando Earthquake, was nothing near as catastrophic.

Geologically, the San Fernando Earthquake was a major seismic event with a magnitude of 6.6 on the Richter scale. During the earthquake, which lasted for 12 seconds, the foothills of the San Gabriel Mountains were shifted several feet vertically and to the southwest along a reverse (thrust) fault. The San Fernando Valley Earthquake of February 9, 1971, was the start of a period of intermittent seismic activity, and is considered to have been the first in a series of damaging earthquakes on reverse faults in the Los Angeles and San Fernando Valley areas that culminated with the 1994 Northridge earthquake.

The faulting that resulted in the San Fernando Earthquake occurred at a depth of eight miles on the Sierra Madre Fault in the foothills of the San Gabriel Mountains. But ground displacement during the earthquake was not limited to the Sierra Madre Fault itself. Ground movement also occurred along a cluster of splinter faults such as the Mission Wells segment, Sylmar segment, Tujunga segment, Foothills segment, and the Veterans fault, all part of what is called the San Fernando Valley Fault Zone. All of the segments shared the common elements of thrust faulting with a component of left-lateral slip, but they were not unified with regard to their connection to the Sierra Madre fault in the underlying basement bedrock.

Widespread surface faulting trending northwest by southeast occurred along the San Fernando Valley Fault Zone. This extended from a point south of Sylmar and stretching nearly continuously for six miles east to the Little Tujunga Canyon. While lateral and vertical faulting were all seen to have occurred during the earthquake, the largest individual component of movement was five feet of left lateral slip near the middle of the Sylmar segment. The largest cumulative amount of slip was six feet along the Sylmar and Tujunga segments. The widespread nature of movement through thousands of feet of loose sand and gravel valley fill expanded the area affected by both shaking and displacement of the ground during the earthquake.

Left-Lateral Movement Across a Section of Highway

Vertical Movement at the Foothill Nursing Home

Damage was locally severe, with the effects being strongest to the south of the epicenter where the earthquake produced extensive surface faulting in the mountains as well as in nearby urban settings. Both the shaking and the horizontal and vertical displacement of the ground affected city streets as well as private homes and businesses. Underground water, sewer, and gas systems did not fare well in the city of San Fernando where the breaks were too numerous to count. Transportation around the San Fernando Valley area was further affected by roadway failures and the collapse of several major freeway interchanges. In places railroad tracks were knocked out of alignment by the earthquake.

The earthquake damaged a number of health care facilities in Sylmar, San Fernando, and other densely populated areas north of Los Angeles. The Olive View Medical Center and Veterans Administration Hospital both sustained very heavy damage, with buildings collapsed at both sites, causing the majority of deaths that occurred during the earthquake. The Foothill Nursing Home, which sat very close to a section of the fault that broke the surface, was raised up three feet relative to the street where surface faulting ran along the sidewalk and across the property.

Section of Railroad Track
Shifted Horizontally by Movement

The near total failure of the Van Norman Dam resulted in the evacuation of tens of thousands residents living in the valley below the dam. Schools were strongly affected, but building codes strengthened after the 1933 Long Beach Earthquake greatly improved the resistance to earthquake damage for the thousands of school buildings in the area. Other results of the earthquake included a methane seep from the floor of the Pacific Ocean near Malibu for several days following the earthquake. In addition, the earthquake caused hundreds of landslides in the San Gabriel Mountains.

The shaking caused by the February 9 earthquake surpassed building code requirements and exceeded what engineers had prepared for. Although most dwellings in the valley were built in the prior two decades, even modern earthquake resistant structures such as the Olive View Medical Center sustained serious damage.

Collapsed Single Story Medical Building
While Popular, the Spanish Tile Roof
Added Weight, Making Collapse
Easier as the Ground Beneath Shifted

A. The Olive View Medical Center

The majority of the buildings at the 880-bed Olive View Medical Center hospital complex had been built prior to the adoption of new construction techniques that had been put in place following the 1933 Long Beach earthquake. Some of the buildings at the large facility escaped damage, but those that did have damage consisted of either wood frame or masonry structures. The five-story reinforced concrete Medical Treatment and Care Building, one of three new additions to the complex, was built with earthquake-resistant construction techniques. It was completed in 1970.

All of the damage to the hospital occurred as a result of shaking caused by the rapidly moving peaks and valleys of the earthquake waves rolling through the ground. Ground motion during the earthquake caused catastrophic damage to the Olive View Medical Treatment and Care building. Three of the four stairwell and elevator towers separated from the main building and fell to the ground, one landing on and smashing the single story hospital administration building. Vertical movement during the passing of the earthquake waves caused many of the support pillars to sheer. In some cases the vertical movement caused the pillars to rupture, exposing the reinforcing steel rods. The entire building was shifted horizontally on its support pillars. The damage to the Medical Treatment and Care Building was such that the structure was subsequently demolished.

All of the emergency vehicles at the Olive View Hospital were parked in a garage. This garage collapsed trapping all of the emergency vehicles.

Stairwell and Elevator Tower
Collapsed on Hospital Administration Building

View of Hospital Building Showing Shift on Support Pillars

Close Up View of Support Pillar
Showing Exposed Reinforcing Steel Bars

View Showing Emergency Vehicles
Trapped in Parking Garage

B. The Veterans Administration Hospital

The Veterans Administration Hospital was originally built as a tuberculosis hospital in 1926; it became a general hospital in the 1960s. By 1971 the facility comprised 45 individual buildings, all lying within three miles of the fault rupture in Sylmar, but the structural damage occurred as a result of the shaking and not from ground displacement. Twenty-six buildings that were built prior to 1933 had been constructed following the local building codes applicable at the time and thus did not meet seismic-resistant design requirements that were implemented after the 1933 earthquake. These buildings suffered the most damage, with four buildings totally collapsing, and resulted in a large loss of life at the facility. Most of the masonry and reinforced concrete buildings constructed after 1933 withstood the shaking and most did not collapse.

View Showing Collapsed Buildings of the
Veterans Administration Hospital

C. The Van Norman Dam

The Van Norman dam was severely damaged as a result of the earthquake. The dam was very close to overflowing and approximately 80,000 people were evacuated for four days while the water height in the reservoir was lowered. This was done as a precaution should further collapse occur due to a strong aftershock.

The damage at the dam consisted of a landslide that dislocated a section of the embankment. The earthen lip of the dam fell into the reservoir and brought with it the concrete lining. What remained of the dam was just five feet above the water level.

The Collapsed Wall of the Van Norman Dam

D. Highways and Interchanges

The San Fernando Valley Earthquake caused substantial damage to about 10 miles of the freeways in the northern San Fernando Valley. Most of the damage occurring at the Foothill Freeway - Golden State Freeway interchange and along a five mile stretch of Interstate 210. On Interstate 5, the most significant damage was between the Newhall Pass interchange on the north end and the I-5 - I-405 interchange in the south, where subsidence at the bridge approaches and cracking and buckling of the roadway made it unusable. The Antelope Valley Freeway had damage from Newhall Pass to the northeast, primarily from settling and horizontal displacement caused by faulting across the roadway. In addition, the earthquake caused splintering and cracking at the Santa Clara River and Solemint bridges. The damage showed just how easily traffic can be disrupted on main thoroughfares, and just how vulnerable we are to damage from an earthquake. And while the magnitude 6.6 San Fernando Earthquake was large, it wasn't anywhere near the strength of earthquakes in the magnitude range of 7 and above.

Collapsed Bridges at Highway Interchange

Collapsed Freeway Bridges

E. <u>Damage to Single Family Homes</u>

Numerous homes in the Sylmar area suffered extensive damage during the San Fernando Earthquake. The shaking of the ground and the vertical movement as the peaks and valley of the rolling earthquake waves passed through the structures caused the damage in most cases. In one case a fault itself passed through a home, causing significant damage.

Split-level houses fared the worst. These houses have a second story built over the garage. Ground movement during the earthquake caused the second story and underlying garage to separate from the rest of the house. Then the weight of the second story caused the garage to collapse, bringing down the whole split-level structure.

Split-Level Second Story Separated from House,
Collapsing the Garage. Note the Car at the Center,
That Had Been Parked in the Garage

When the Shaking Stopped These Two Houses
Were Swaying in Opposite Directions,
Ending the Earthquake Leaning Towards One Another

New Construction Without Lateral Support

New Construction Completely Collapsed

Ground Displacement Along a Fault
Went Right Through This House

3. Oklahoma:
Earthquake Capitol Of The United States

Earthquake Damage to Highway 62
Near Prague, Oklahoma, November 5, 2011

I was born and raised in Southern California, historically the center of earthquake activity in the contiguous United States. I experienced a lot of earthquakes during 34 years living in Southern California, mostly small, but some that were sizeable. When my wife Kathy and I moved to Oklahoma in 1979 I thought I had left all of the rumbling and shaking behind. And I had, for 32 years.

I am a geologist, retired from Kerr-McGee Corporation. My wife passed away in 2001, and since 2006 I have been living with family in Cushing, Oklahoma. My introduction to the Cushing area began in 1984 when I first became involved in the hydrogeologic investigation, and subsequent cleanup, of a former Kerr-McGee refinery site on Deep Rock Road north of town. I was intermittently involved with that project for 17 years. Cushing is a very nice community full of wonderful people and amenities and, more recently, earthquakes.

Oklahoma has experienced a dramatic rise in the frequency and severity of earthquakes in recent years. A state that experienced an average of less than two earthquakes a year with a magnitude of 3 or higher during the previous thirty or more years was suddenly being jostled with hundreds in recent years. Thousands of earthquakes of all sizes have occurred in Oklahoma and surrounding areas in southern Kansas since 2009.

While I was curious about the rise in earthquake activity in normally benign Oklahoma, for the most part I ignored them. I was more anxious about tornadoes; small earthquakes were not very significant to someone who grew up in Southern California.

Then came THE BIG ONE on November 5, 2011. My vacation from earthquakes ended on that day, the date of the Prague Earthquake.

For several days prior to that November day there had been a swarm of small earthquakes reported in the area between Prague and Sparks, located 40 miles east of Oklahoma City in Lincoln County. Then, on the early morning of November 5, 2011, a notable earthquake with a magnitude of 4.8 struck the area. Later that same day, at 10:53 pm., an earthquake with a magnitude of 5.6 rocked the same area.

I was lying in bed in my family's Cushing home when I heard what sounded like the rumble of an approaching freight train, increasing in intensity until the ground started shaking. And what shaking it was! This old farmhouse creaked and groaned as the wood structure twisted and swayed. And then THE BIG One was over.

But that didn't end the earthquake activity. The swarm of earthquakes continued for weeks, punctuated by a 4.7 magnitude aftershock on November 7. Through the end of 2011 a total of 64 more earthquakes were recorded in the same area, nearly double the number recorded in all of 2010. After it was all said and done the United States Geological Survey (USGS) designated the magnitude 4.8 earthquake early on November 5 a "foreshock," with the magnitude 5.6 earthquake being called the "mainshock."

What became known as the Prague Earthquake, the magnitude 5.6 earthquake was the largest earthquake ever recorded in Oklahoma. The previous record was a magnitude 5.5 earthquake that struck near the town of El Reno, located west of Oklahoma City, on April 9, 1952, 59 years earlier.

The Prague Earthquake resulted in considerable damage in a wide area. Portions of Highway 62 near the epicenter were fractured and displaced by earth movements in three locations. Several nearby homes suffered damage due to collapsed brick walls and fallen chimneys. One chimney fell through the roof of a house, doing considerable damage. In Shawnee, in Pottawatomie County about 20 miles to the south of Prague, one turret at St. Gregory's University collapsed completely, with two other turrets sustaining damage.

Damage to Highway 62 Resulting from the
Prague Earthquake

Throughout a larger area there were cracked walls and lots of broken windows and items knocked off shelves. The home of a friend of mine was made uninhabitable by structural damage resulting from the earthquake. And, understandably, there were a lot of frayed nerves.

Homeowner's Insurance doesn't cover earthquake damage, and very few people had earthquake policies. Who needs earthquake insurance in Oklahoma? Even St. Gregory's University didn't have earthquake insurance.

Collapsed Chimney Resulting from the Prague Earthquake

Damage to Home from Collapsed Chimney

There was understandable media frenzy after the Prague Earthquake. While there had been a lot of buzz about all of the small earthquakes Oklahoma had experienced during the two years leading up to the Prague Earthquake, talk about earthquakes suddenly took center stage. As the swarm of small earthquakes continued in the Prague area, as well as in other parts of Oklahoma, the local television stations began jokingly giving earthquake forecasts along with weather forecasts. Some electronic highway billboards displayed the magnitudes of recent earthquakes. It was becoming something of a circus.

Cracked Wall and Broken Glass
Resulting from the Prague Earthquake

Repairing Damage to Collapsed Turret at
St. Gregory's University

But all of the earthquakes that were happening were no laughing matter. People's lives were being disrupted, homes damaged, knickknacks knocked off of shelves and dishes out of cupboards; homeowners were left to pay to repair the damage themselves.

And there was the fear of even bigger earthquakes, bigger than that in Prague.

Oklahoma is crisscrossed by hundreds of miles of oil and natural gas pipelines. Cushing itself is referred to as the "Pipeline Crossroads of the World" because so many pipelines converge there. And there are hundreds of huge oil storage tanks in the Cushing area to accommodate all of the oil moving through on its way to refineries. Could you imagine the damage that would happen from ground rupture like what damaged Highway 62 should it happen under at a pipeline or a storage tank? Or should it happen under a school?

Part of a Cushing Oil Storage Tank Farm

The Prague earthquake got my attention. It piqued my interest as a geologist to know why an area in the middle of the North American continent, far away from any geologically active tectonic plate boundaries, should suddenly be experiencing so many earthquakes. My journey of understanding was both informative and shocking.

A. A Word About Earthquake Magnitudes

The Richter scale of earthquake magnitudes is not linear, but rather logarithmic. An earthquake with a magnitude of 5 produces shaking that is ten times stronger than one with a magnitude of 4, which is ten times stronger than a 3. An earthquake with a magnitude of 5 thus produces shaking that is 100 times stronger than one with a magnitude of 3. So when I write about a foreshock with a magnitude of 4.8 and a mainshock of 5.6, the energy released by the Prague Earthquake mainshock was substantially greater than the foreshock.

While an earthquake in the magnitude 2 range can sometimes be felt, those of magnitude 3 and above are generally more noticeable. Earthquakes in the range of magnitude 3 and higher will herein be termed "significant."

B. What Could Have Caused the Prague Earthquake?

Everyone wanted to know why, in a state not known for seismic activity; there were suddenly so many significant and even damaging earthquakes. As a retired geologist I wanted to know why we were having so many noticeable earthquakes in an area where there should be few, if any, such earthquakes. And certainly nothing on the scale of the Prague Earthquake.

What was immediately labeled the probable culprit by the media was "fracking."

To produce oil and natural gas from "tight" rocks such as shale and limestone, it is often necessary to increase the openness of the pore spaces or create fractures in the rock so that the oil and gas can flow through the rock to the production well. A process of hydraulic fracturing, otherwise known as "fracking," usually does this.

Fracking is a method used to enhance the recovery of oil and gas from wells. It involves tapping shale and other tight-rock formations by drilling a mile or more below the surface before gradually turning horizontal and continuing several thousand feet more. Several wells can be constructed at one drilling site using this process. After construction, a mixture of mostly water, with some sand and chemical additives, is pumped into the well at high pressure to create micro-fractures in the rock. The small fractures are held open by the sand grains, boosting well efficiency. Older wells can also be subjected to fracking to enhance production.

Fracking is nothing new, but is a process that has been in use for more than 65 years. What is new is the number of wells being constructed using fracking. The combination of advanced hydraulic fracturing and horizontal drilling technologies used today is mostly responsible for the recently surging oil and natural gas production in the United States. The use of fracking technology is the biggest single reason America is having an energy boom right now, one that has changed the energy picture in the United States from one of scarcity to abundance. Fracking is letting the United States tap vast oil and natural gas reserves that previously were locked away in tight-rock formations.

I personally support fracking; I am in favor of anything, within reason, that helps make the United States energy independent.

Fracking is NOT causing most of the significant induced earthquakes we have seen in Oklahoma in recent years. Rather, wastewater disposal is the primary cause of the recent increase in earthquakes. Wastewater disposal wells typically operate for longer durations at much greater depths and inject much more fluid than hydraulic fracturing, making them more likely to induce significant earthquakes.

Wastewater, mostly saltwater, is produced from all oil wells, not just fracking sites. The process of fracking itself produces only a small percentage of the wastewater from a well that must be disposed of in a wastewater disposal well.

Most wastewater currently disposed of in Oklahoma is saltwater produced in the process of oil and gas extraction. This is saltwater that was locked up in the sediments when they were first deposited under the ocean millions of years ago This saltwater is a byproduct of the oil and gas extraction process, and is found at nearly every oil and gas extraction well.

Most of the saltwater produced in north central and central Oklahoma comes from two main rock units. These are the Mississippi Lime unit in north-central Oklahoma and extending into Kansas, and the Hunton unit in central Oklahoma. Both of these oil- and gas-producing units, which occur at depths of 3,000 to 6,000 feet, produce large volumes of saltwater along with the petroleum. Over time there can literally be millions of gallons of saltwater produced with oil from a single well that must be eliminated in an environmentally conscious manner.

It has been estimated that oil and gas wells in Oklahoma produce ten barrels of saltwater for every barrel of oil. Since a barrel is 42 gallons, that is 420 gallons of saltwater for every 42 gallons of oil. That is a lot of saltwater.

In past years some operators commonly disposed of saltwater produced from oil wells in ponds and streams. This had an obviously negative impact on the local aquatic ecology. A better solution had to be found, and that is where the deep wastewater disposal wells come in. Most of this produced saltwater is now disposed through injection into deep wastewater disposal wells. These wells are drilled much deeper than oil wells.

Disposal of all of this saltwater through deep wastewater injection wells is the real culprit. The USGS, along with the University of Oklahoma and Columbia University determined the source of the Prague Earthquake was active wastewater disposal wells nearby.

So fracking has got a bad rap; fracking itself is not the cause of the Prague Earthquake nor any of the other significant earthquakes that have shook Oklahoma in recent years. The process of fracking can cause extremely small earthquakes, but these are much too small to be felt.

Neither is enhanced oil recovery projects employing water flooding causing the earthquakes. Water flooding involves the injection of fluid into rock layers where oil and gas have already been extracted to force out more oil and gas. While water flooding can cause small earthquakes, they are generally too small to be felt.

Wastewater disposal wells generally inject fluids into much deeper rocks not tapped for any other purpose, at greater depths where significant earthquake causing faults are likely to be found.

C. It's All About the Geology

The geology of central Oklahoma consists of a thick sequence of sedimentary rocks, 8,000 feet or more in thickness, underlain by crystalline rock, mostly granite, of the Earth's crust. This crystalline rock is referred to as the "basement." All of the oil and gas produced in Oklahoma is found in the thick sequence of sedimentary rock. As noted above the Mississippi Lime and the Hutton oil and gas producing units are found at depths between 3,000 and 6,000 feet.

In central Oklahoma many wastewater disposal wells are completed in the Arbuckle Formation, a sedimentary rock unit found at a depth of 7,000 feet or more in central Oklahoma. The Arbuckle Formation is a favorite for injecting wastewater because it is "under pressurized," making it easier to inject fluids. The problem is that, because the Arbuckle Formation sites right on top of the crystalline basement rock, disposing of wastewater in the Arbuckle Formation can result in notable earthquakes. It is all about the geology.

There is a reason why the epicenters of so many significant earthquakes in central Oklahoma are typically reported at depths of three to five miles. The faults along which the earthquakes occur have their roots deep in the crystalline basement rocks. Depending on the faults impacted by wastewater disposal, induced earthquakes can occur at significant distances from the sites of wastewater disposal wells and at greater depths.

The Oklahoma Geological Survey (OGS) has determined that the Prague Earthquake of November 5, 2011, occurred along the Wilzetta Fault, also known as the Seminole Uplift. The Wilzetta Fault is a 55-mile long fault zone that runs from central Pottawatomie County south of Prague, through Lincoln County, to the western part of Creek County to the north. It is a strike-slip fault, moving two adjacent blocks of the crystalline basement rock of the Earth's crust horizontally past one another, similar to the San

Andreas Fault in California. Unlike the San Andreas Fault, though, central Oklahoma in the middle of the North American continent far from any tectonic plate boundaries.

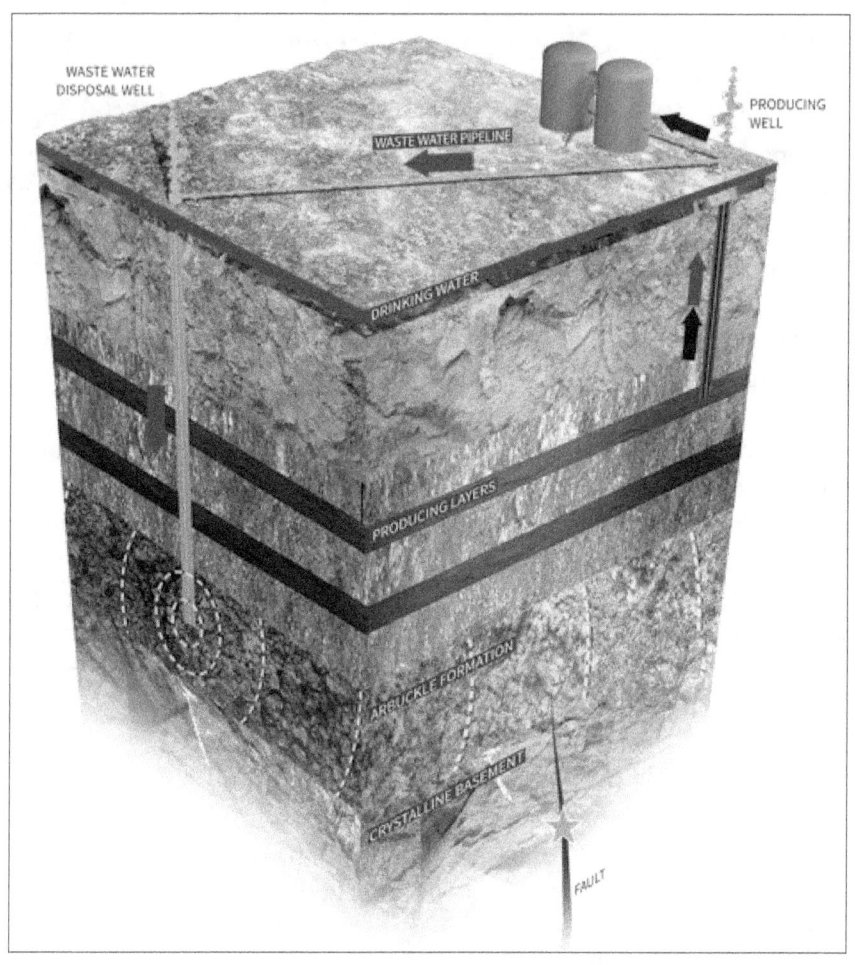

Simplified Geologic Profile Showing Oil Producing Wells and Wastewater Disposal Wells

The Wilzetta Fault, as well as other basement faults in central Oklahoma, is the result of tectonic forces created tens of millions of years ago. During the late Paleozoic Era, 250 to 300 million years ago, the crystalline basement rocks that comprise the Earth's crust of North America were near two tectonic plate boundaries. A super continent called Pangea was being created as the African continent slowly collided with the east coast of North America, creating the Appalachian Mountains in the process. During the same time the

South American continent was colliding with the south coast of North America, creating the Ouachita, Arbuckle and Wichita Mountains in southern Oklahoma. These collisions between continents produced tremendous compressional stresses in the crystalline basement rocks of North America, resulting in the creation of fractures in the rock, along which blocks of crust can move relative to one another.

Millions of years later the super continent of Pangea itself broke apart as continental drift proceeded. What became Europe and Africa split from North America to produce the Atlantic Ocean. The South American continent split from North America and moved south to create the Caribbean Sea.

The basement rock in Oklahoma is still rebounding from the stresses created so long ago. All it takes is something to stimulate, or lubricate, the fault zones to release pent-up stresses in the crystalline basement rock. Research done during the past fifty years has shown that deep wastewater disposal wells can provide the necessary stimulus to cause movements along the basement faults, resulting in earthquakes.

The Arbuckle Formation rests right on top of the fractured crystalline basement rock. Injected wastewater can penetrate already stressed basement faults. This wastewater increases the pore pressure in the Arbuckle Formation and the underlying basement rock. This increase in pore pressure can open and lubricate fracture planes leading to slippage along fault lines in the underlying crystalline basement rock, releasing pent-up stress. And the closer the bottom of the wastewater disposal well is to the crystalline basement the more likely wastewater disposal is to induce earthquakes. Some of these earthquakes are large enough to be felt; some may be strong enough to cause damage.

There have been cases where wastewater disposal wells have been drilled into the crystalline basement rock. This is a worst-case scenario for causing earthquakes, and the State of Oklahoma is working to eliminate the problem.

It takes time for injected wastewater to move through the Arbuckle Formation and into the underlying crystalline basement rock. For that reason, there may be a significant time delay from when wastewater disposal begins and earthquakes start. Once the wastewater gets into the crystalline basement rock, though, it can travel quickly over long distances and to greater depths. For that reason, earthquakes can occur on faults some distance and depth from the well. And because so much wastewater in injected in disposal wells and so much pressure is built up, it can take years for induced earthquake activity to end after injection stops.

It must be noted that not all wastewater injection wells induce earthquakes; most injection wells are not associated with felt earthquakes. Many factors are necessary for wastewater disposal to cause significant earthquakes. These include the depth of the well, how close the bottom of the well is to the crystalline basement rock, the injection rate and total volume injected, and the presence of faults in the crystalline basement that are large enough to produce significant earthquakes. These faults must have stresses built up that are large enough to produce significant earthquakes. And there must be pathways present for the fluid to travel from the injection point to the faults. In the case of the Prague Earthquake all of these factors were present to produce the magnitude 5.6 event.

D. Oklahoma: The Center of Seismic Activity in the United States

Beginning in 2009, the frequency of earthquakes in Oklahoma rapidly increased from an average of less than two magnitude 3 and higher earthquakes per year to hundreds in 2014 and 2015. A state that historically had experienced few earthquakes was suddenly being jostled with thousands in recent years, as has surrounding areas in southern Kansas.

In 2014, there were three times as many earthquakes recorded in Oklahoma than in California, making Oklahoma the most seismically active state in the contiguous United States. When looking only at significant earthquakes, those with a magnitude of 3 or higher, ones that are likely to be felt, in 2015 there were six times as many significant earthquakes in Oklahoma than in California. To make matters worse, seismic activity in Oklahoma has been spreading northward into Kansas, which experienced a major jump in earthquakes in 2013 and 2014.

A map of earthquake epicenters in Oklahoma shows two main areas of activity, central and north-central Oklahoma. These groupings correspond to the areas where most of the oil and gas recovery operations are taking place, and thus where most of the wastewater disposal wells are located. The trend is northwest by southwest following the trend of the oil and gas-producing units.

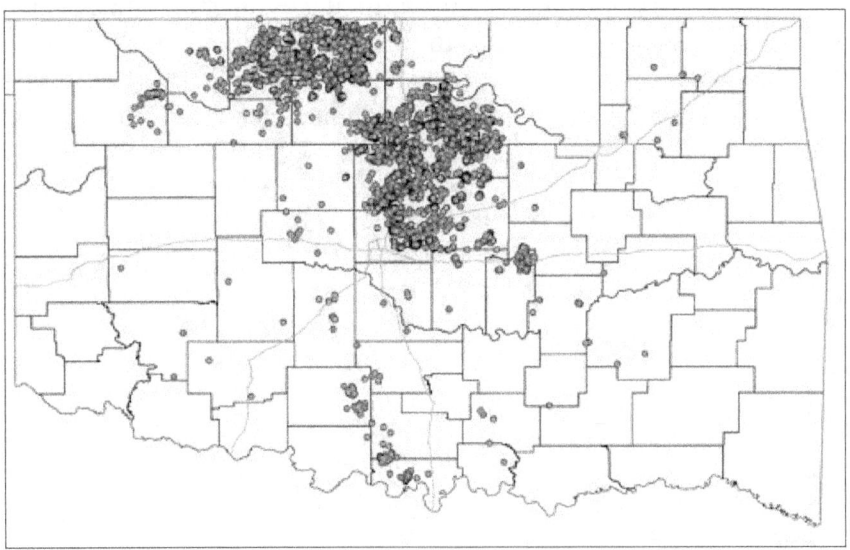

Map Showing Earthquake Epicenters in Oklahoma

The total number of earthquakes has been increasing dramatically in Oklahoma in recent years. During the period 1970 through 2009 there were a total of 1,861 earthquakes of all sizes recorded in Oklahoma. In 2010 the total number of earthquakes was 1,060; in 2011 the total number was 1,542. The total number of earthquakes recorded in 2012 dropped to 1,027, but rose to 2,848 in 2013, more than had been recorded in the period 1970 through 2009! The total number of earthquakes rose dramatically to 5,399 in 2014, increasing to 5,690 in 2015.

Most of these earthquakes were too small to be felt. While earthquakes in the rage of magnitude 2 can be felt by some, earthquakes with a magnitude of 3 are generally considered to be the most widely felt.

During the thirty-year period 1978 through 2008, Oklahoma experienced an average of less than two earthquakes a year with a magnitude of 3 or higher. In 2009 that number had jumped to 20, doubling to 42 in 2010. The number increased again to 62 in 2011, but dropped to 55 in 2012.

After 2012 the number of earthquakes of magnitude 3 and higher really began to increase. In 2013 the number had doubled to 109. The number increased dramatically to 585 in 2014, then to 907 in 2015. Oklahoma is definitely the earthquake capitol of the United States!

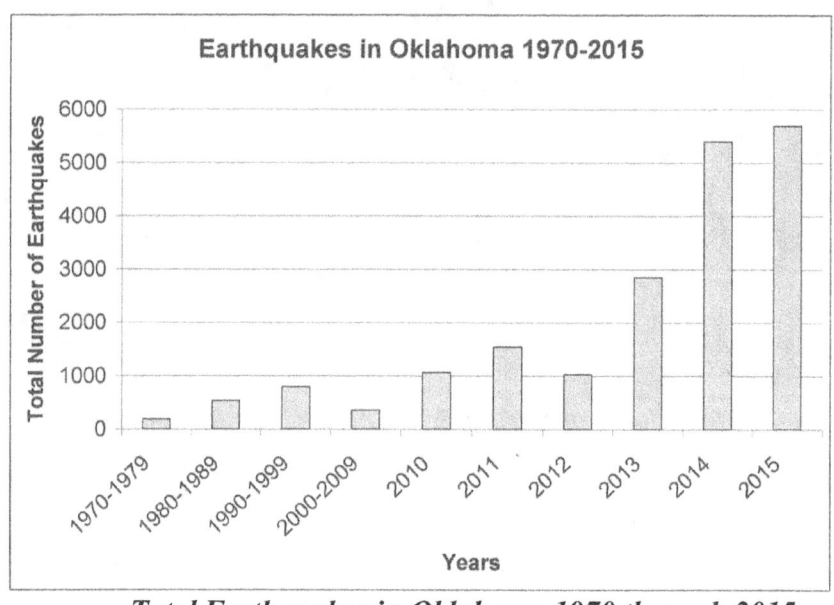

Total Earthquakes in Oklahoma 1970 through 2015

E. The Evidence Regarding Man-Made Earthquakes

At the time of the Prague Earthquakes of November 5, 2011, there were 187 wastewater disposal wells operating in Lincoln County alone. As of February 2016 there were more than 4,000 injection wells operating in Oklahoma, of which 1,000 were injecting fluids into the Arbuckle Formation.

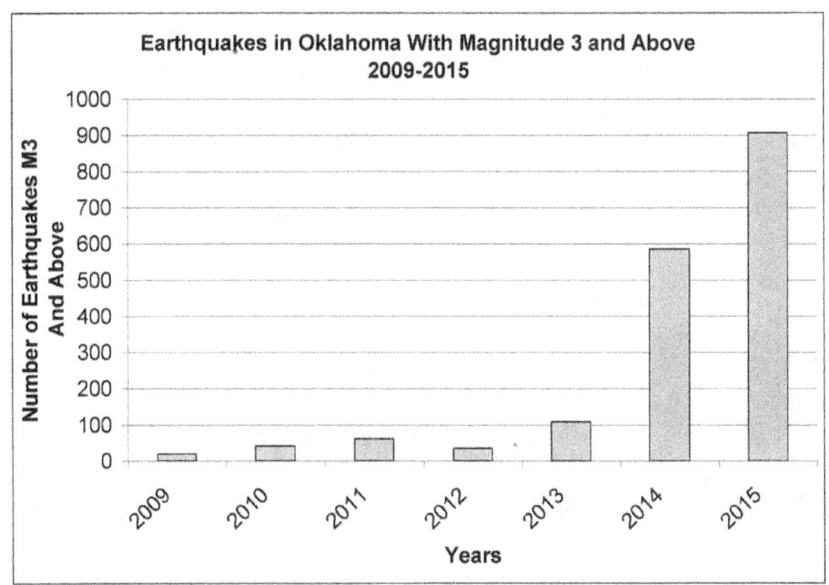

Earthquakes with a Magnitude of 3 or Higher

In 2013 a seismologist at the University of Oklahoma published a paper regarding the relationship of wastewater disposal wells and the November 5, 2011, Prague Earthquake, concluding there was a "compelling link between the zone of injection and the seismicity." The Oklahoma Geological Survey (OGS), however, rejected these findings, stating, "The interpretation that best fits current data is that the Prague earthquake sequence was the result of natural causes."

Regardless of the OGS's conclusion, scientists, especially those at the USGS, have known since at least the 1960s that wastewater injection, during which millions of gallons of wastewater are injected into a disposal well, can induce earthquakes by increasing the fluid pressures underground. That increasing pressure reduces the frictional strength of faults, allowing them to slip, causing earthquakes.

The first case study demonstrating the cause and affect relationship between wastewater disposal wells and induced earthquakes involved the Rocky Mountain Arsenal located northeast of Denver, Colorado. In 1961 the Army Corps of Engineers drilled a 12,000-foot well at the facility to dispose of waste fluids from

Arsenal operations. These fluids consisted of water that included some metals, chlorides, wastewater and toxic organics. The primary contaminants were organochloride pesticides, organophosphate pesticides, carbamate insecticides, organic solvents such as chlorinated benzenes, heavy metals, chemical warfare material and their related breakdown products and biological warfare agents.

The injection of wastewater was started in March 1962. Shortly thereafter an unusual series of earthquakes began. The first shock came on April 24. By the end of December 1962, 190 earthquakes had occurred. Several were felt, but none caused damage until a stronger earthquake occurred on December 4. Between January 1963 and August of 1967 more than 1,300 earthquakes were recorded in the area, which increased in intensity as time went by.

Three earthquakes in 1965 caused noticeable damage in the area. Then, in October 1966, a stronger earthquake rocked a 15,000 square-mile area of Colorado. This earthquake caused damage in the form of broken windows and dishes, and cracked walls and plaster. Another strong earthquake shook the Denver area on November 14, 1966, causing more damage. Smaller earthquakes continued throughout the remainder of 1966 and through the first week of April 1967.

Then, on April 10, 1967, the largest earthquake since the swarm began in 1962 occurred. This earthquake, which had a magnitude of 5.0, broke windowpanes and caused cracks in asphalt roadways. Some schools in Boulder, Colorado were closed because of cracks in walls.

The strongest and most widely felt shock in Denver's history struck the area on August 9, 1967. The magnitude 5.3 tremor caused the most serious damage of any of the earthquakes. Concrete pillar supports to a church roof were weakened. Many homeowners reported wall, ceiling, floor, patio, sidewalk, and foundation cracks. Several homeowners reported basement floors separated from walls. The earthquake caused a million dollars in damage. An aftershock with a magnitude of 4.4 occurred on April 27, 1967.

During November 1967, several significant earthquakes shook the Denver area. A magnitude 4.1 occurred on November 15. Then the second largest earthquake in Denver history occurred on November 26, 1967. The magnitude 5.2 event caused widespread minor damage in the suburban areas northeast of Denver. Merchandise fell from supermarket shelves and walls cracked in some larger buildings. It was felt at Laramie, Wyoming to the northwest, Goodland, Kansas to the east, and south to Pueblo, Colorado. And all of this induced earthquakes was coming from just one wastewater disposal well!

The injection of wastewater into the Army's disposal well at the Rocky Mountain Arsenal was discontinued in 1966 because of the suspicion that it was causing the earthquakes that were affecting the Denver area. In September of 1968 the Army began removing fluid from the waste disposal well at a very slow rate, in the hope that earthquake activity would lessen, which it did. By ten years later, in 1976, the earthquakes had ended for the most part.

As seen with the Rocky Mountain Arsenal wastewater disposal well, there is typically a time delay between peak wastewater injection rate and the onset of seismic activity. There is also a spatial separation between the epicenter of the earthquakes and the disposal well sites. Some earthquakes can occur months or even years after injection rates have peaked, and in locations that were sometimes located miles away from the well.

Wastewater increases the pore pressure in the disposal zone, in the Rocky Mountain case in the crystalline granitic basement rock, and in Oklahoma case in the Arbuckle Formation and the underlying crystalline basement; the crystalline basement rock is from where most of the significant earthquakes originate. Because pressure from wastewater injection is spreading throughout the formation, it can affect faults located far from the well site, creating a time delay between the time wastewater disposal begins and the onset of movement along a fault. If a fault weren't located directly beneath a well, but several miles away, it would take time for the fluid pressure to propagate to the fault zone.

Following the discovery in 1966 that injection of fluid underground at high pressure at the Rocky Mountain Arsenal was responsible for the triggering of earthquakes near Denver, the USGS speculated that induced earthquakes might be controllable. Reduction of the frictional strength of the highly stressed basement rock by injection of the fluid is the favored explanation for the mechanism by which the induced earthquakes were triggered.

The USGS decided to run a field experiment to see if earthquakes caused by wastewater disposal wells could indeed be controlled. Earthquakes at the Rocky Mountain Arsenal were caused by injection of wastewater into stressed rock. When the fluid pressure in the Rocky Mountain Arsenal well was reduced the earthquakes sharply decreased in frequency. The USGS believed that if the physical basis for the phenomena could be well established in a field experiment, earthquake control and prevention of inadvertent triggering of earthquakes might become feasible.

The disposal well at the Rocky Mountain Arsenal, because of its proximity to Denver, could not be used for experimental purposes; the USGS didn't want to create any more earthquakes in the Denver area. A second site with an active injection well was selected at the Rangely Oil Field in Colorado. An array of seismographs located nearby at Vernal, Utah, had been recording small earthquakes from the vicinity of Rangely Field since installation of the seismographs in 1962. Water flooding at high pressure for secondary oil recovery had been taking place in the Rangely Field since 1957.

In the fall of 1967 the USGS installed an array of portable seismographs, after which 40 small earthquakes were recorded during a 10-day period at the oil field. The earthquakes occurred in two areas within the oil field where fluid pressures due to water flooding were high. All of these earthquakes were too small to be felt, but were detected by seismographs.

After a year of recording of seismic activity in the area, the fluid pressure in the vicinity of the earthquakes was first reduced by backflowing water from injection wells to the surface. This drop in fluid pressure resulted in reduced seismic activity. The pressure was then increased again by injection and the cycle repeated, resulting in a corresponding increase in seismic activity.

Over time the USGS scientists found a direct correlation between increased seismicity and high pore pressure. The experiment at the Rangely field confirmed the hypothesis that earthquakes may be triggered by increases in fluid pressure. The strong correlation between frequency of the seismic activity and variations in the fluid pressure around the predicted value is evidence for this conclusion.

Years later, in a July 2001 study, "Technical Program Overview: Underground Injection Control Regulations" the United States Environmental Protection Agency (USEPA) agreed with the USGS that the deep, hazardous waste disposal well at the Rocky Mountain Arsenal had causing significant seismic events in the vicinity of Denver, Colorado."

Another case study involved disposal of oil field saltwater in a deep well in 1991. In the late 1980s the United State Bureau of Land Management proposed the disposal of saltwater produced from oil field operations in a deep limestone unit to prevent the water from entering a river in the Paradox Valley of Colorado. Taking advantage of the proposal to further quantify the cause and affect relationship of wastewater disposal and earthquakes, scientists started monitoring earthquake activity before the wastewater disposal project started; there was no significant seismic activity in the area. After injection of wastewater in the disposal well began in 1991, more than 6,200 earthquakes were recorded.

Since then the USGS has been involved in many projects regarding wastewater disposal wells and earthquakes. In 1990 the USGS submitted a report to the USEPA titled "Earthquake Hazard Associated With Deep Well Injection." This report chronicled case histories of earthquakes associated with well operations. In 1992 Craig Nicholson and Robert Wesson published a report titled "Triggered Earthquakes and Deep Well Activities."

The volume of wastewater injected in deep disposal wells in Oklahoma tells the cause and affect story. In 1997, about 20 million barrels of produced saltwater were injected in just three areas of Oklahoma. In 2012 that number was up to 112 million barrels. The total was 400 million barrels in 2013. In 2014 a total of more than 750 million barrels of wastewater had been injected into wastewater disposal wells in just the top ten producing counties in Oklahoma.

In Payne County, where I live, a total of more than 72 million barrels of wastewater were injected in wastewater disposal wells in 2014. That is 3,024,000,000 gallons of wastewater, an increase of 479% over what had been injected in wastewater disposal wells in 2012. This increase in wastewater disposal has significantly increased the number and magnitude of earthquakes in Payne County.

F. Government Response

A state that gets a lot of its revenue from the oil and gas industry was understandably slow to acknowledge that industry operations were causing earthquakes. That began to change in 2010.

It has been since at least 2013 that the Oklahoma Geological Survey (OGS) has acknowledged a potential link between oil and gas operations and the ongoing earthquakes. In that year the OGS stated that the agency had recognized the potential for the connection between a swarm of earthquakes in 2010 near Jones, east of Oklahoma City, to be due to the oil and gas operations.

In October 2013, a joint statement between the OGS and the USGS announced that "activities such as wastewater disposal" might be a "contributing factor to the increase in earthquakes."

In September 2014, in response to the increasingly noticeable swarms of earthquakes, Oklahoma Governor Mary Fallin announced the creation of a "Coordinating Council on Seismic Activity" to help promote further understanding of the increase in seismicity. Governor Fallin added, "We believe that by linking scientists and energy experts, we can develop sound regulatory practices and policies in our state while also alleviating any questions our citizens might have."

Faced with insurmountable scientific evidence, on April 21, 2015, the OGS said in an official statement that it "considers it very likely that the majority of recent earthquakes, particularly those in central and north-central Oklahoma, are triggered by the injection of produced water in disposal wells." The OGS added that in 2013 the seismicity rate in Oklahoma was 600 times greater than the background seismicity rate observed in Oklahoma prior to 2008. The OGS added that this increase in seismicity was "...very unlikely the result of a natural process."

It wasn't long after the comments by the OGS that people began to sue the energy industry for damages suffered in "man-made" earthquakes. On June 30, 2015, the Oklahoma Supreme Court ruled that homeowners who had suffered injuries or property damage as a result of frequent earthquakes believed to be caused by industrial activities could legally sue for damages in trial courts. This ruling was issued in spite of efforts by the industry to prevent such lawsuits from being accepted.

A person who had sustained injuries during the 2011 Prague Earthquake originally brought the case before a state court. At that time the court denied the claim for damages. After the OGS admitted that the earthquake was caused by energy industry activities, the case was appealed to Oklahoma State Supreme Court, which found in the plantiff's favor.

In August 2015 Oklahoma Governor Mary Fallin and State Secretary of Energy and Environment Michael Teague met with members of the Coordinating Council on Seismic Activity to talk about earthquake activity in the state. The purpose of the meeting was to brief the governor on increased seismic activity, the cause, and what can be done. During the meeting the governor acknowledged there is a direct correlation between increased seismic activity and disposal wells, adding that fracking wasn't causing the earthquakes, but rather that wastewater disposal was the cause.

The governor added that the council had already taken several steps to address the disposal well problem. In March 2015, 347 wastewater disposal wells were ordered to reduce injection rates. In the following July, 211 wells additional were ordered to do the same. The governor added that, because it takes time for wastewater to move through the rock, it could be at least a year before we see any measurable reduction in the number of earthquakes.

To address the problem, the OGS developed a "traffic light" system, which assigns "yellow light" status to some disposal wells when earthquakes occur, imposing limits on injection rates and pressures. The system assigns "red light" status to other injection wells, requiring them to shut down completely when earthquakes occur. The state also ordered that more than 50 disposal wells drilled down into the crystalline granitic basement rock to be plugged back to more shallow depths to keep injected wastewater from entering fractures in the crystalline basement rock.

The traffic light system seems to be working. Following earthquake activity in Alfalfa County in north-central Oklahoma near the Kansas border in late-January 2015, SandRidge Energy was ordered to shut down an injection well it was operating. This was the second wastewater injection well directed to halt operations by the agency since the new monitoring system was established. The directive was issued on February 3, 2015 in response to a magnitude 4.1 earthquake recorded in the area four days earlier. SandRidge Energy was operating under a 'red light' permit with language that said shut down if there's any seismic activity.

But is the state's actions enough? How do we protect the public while letting the energy industry operate? The energy industry, after all, is a very big part of the economy in Oklahoma and a necessity to make the United States more energy independent.

I personally have noticed a decline in the number of earthquakes felt the Cushing area. It seems that perhaps the state's traffic light system is having some affect. But not all areas have been so lucky.

On December 29, 2015 a magnitude 4.3 earthquake struck near tbe town of Edmond in northeastern Oklahoma County. This was the 29th earthquake of magnitude 4 or greater to shake Oklahoma in 2015. That earthquake was followed ten minutes later by a magnitude 3.4 aftershock centered nearby. The magnitude 4.3

earthquake caused structural damage to at least one home and caused power outages to over four thousand residents.

The New Year 2016 started off with a good shake. A magnitude 4.2 earthquake struck on the morning of January 1, 2016, again in northeastern Oklahoma County. Then on January 6, two earthquakes, one of magnitude 4.4 and one of magnitude 4.8, struck within less than a minute of each other in north-central Oklahoma.

Then, on the morning of February 16, 2016, the city of Fairview in northwest Oklahoma was struck by a magnitude 5.1 earthquake, the third strongest quake ever recorded in the state. I felt that earthquake in Cushing. It was followed by several aftershocks during the next 90 minutes, including one with a magnitude of 3.9.

For that reason the Oklahoma Corporation Commission (OCC) has recently become even more proactive in its approach to the problem. On February 16, 2016, the same day as the magnitude 5.1 earthquake, the OCC implemented the largest volume reduction plan yet for wastewater disposal wells in western Oklahoma. Stating that earthquake activity in the region demanded a regional response, the plan covered 245 disposal wells injecting wastewater into the Arbuckle Formation. In conjunction with the 191,000 barrel a day reduction plan begun in northwest Oklahoma earlier in the year, the total volume reduction will be more than half a million barrels a day, or about 40 percent.

Despite these actions, Oklahoma Geological Survey data clearly showed the need for a larger, regional response. That is why, even as the OCC took actions in various parts of the region in response to specific earthquake events, it was already working on a larger plan. That larger plan will include wastewater disposal wells located in areas that are not yet experiencing major earthquakes. The OCC's goal is aimed not only at taking further action in response to past activity, but also to get out ahead of it and hopefully prevent earthquakes in new areas.

Some provisions of that larger plan were announced on March 7, 2016. On that day the OCC announced actions to further reduce the amount of wastewater injected through more than 400 wells completed in the Arbuckle Formation in areas where earthquakes have occurred. The purpose of this action was to reduce the total volume of wastewater injected to 40 percent below the 2014 total. This amounts to a reduction of more than 300,000 barrels a day from the 2015 average injection volumes.

On the same day the OCC added more restrictions on disposal well operations in areas that have not yet seen major earthquake activity. This action affected 118 more wells completed in the Arbuckle Formation.

The new actions also require wastewater disposal well operators to prove that their wells have not been drilled too deep, into the crystalline basement rock. Wells completed in the crystalline basement inject their wastewater directly into the basement fractures and associated faults, which can result in significant earthquakes. Wells that are drilled too deep must be plugged back to a depth within the Arbuckle Formation.

The March 7 actions also eliminated the possibility of administrative approval of new wastewater disposal wells completed in the Arbuckle Formation. This is the OCC's proactive move to get ahead of the earthquake activity.

It remains to be seen how effective the state's traffic light system will be. Because it takes a long time for pressurized wastewater to find its way to a fault, maybe it is too early to celebrate. As seen from the Rocky Mountain Arsenal wastewater disposal well, which was shut down in 1966, because of the volume of wastewater already injected in deep disposal wells it could be a few years before earthquakes in Oklahoma are reduced in number.

As of this writing in May 2016 Oklahoma has seen more than 3,100 measurable earthquakes in the last year. As of May 18, 2016, a total of 101 earthquakes have been recorded during the last month, and 19 during the last week. The largest earthquake recorded in Oklahoma in 2016 continues to be the magnitude 5.1 event recorded in Fairview on February 16. During the last week the community of Luther, Oklahoma was hit by a magnitude 3.9 earthquake.

Because so many induced earthquakes have struck Oklahoma in recent years the USGS is, for the first time, including earthquakes believed to have connections to industrial activities in its National Seismic Hazard Map. This maps sets standards for construction and insurance rates. The new map became available at the end of 2015. Because all of these earthquakes are now considered to be caused by human activity, earthquake insurance is no longer available to residents.

G. July 2016 Update

Since this report was written in May 2016, the Oklahoma Corporation Commission's proactive efforts to reduce the number of induced earthquakes have had some success. I have not noticed any significant earthquakes in the Cushing area itself. However, during the one-year period since July 2015 there have been a total of 2,701 earthquakes recorded in Oklahoma.

And as of July 31, 2016 the magnitude 5.1 earthquake that occurred in Fairview in northwestern Oklahoma on February 16, 2016 remains the largest to occur during the year. But Oklahoma continues to shake on a regular basis. An earthquake with a magnitude of 3 or higher occurs almost every day, and during many weeks there is at least one earthquake with a magnitude of 4 or higher. Most of these very noticeable earthquakes have occurred in north central and northwestern Oklahoma, with the Fairview area continuing to be a center of earthquake activity.

There were a total of 111 earthquakes recorded during the month of July 2016, the largest of which was a magnitude 4.4 that occurred in early July in the Fairview area. The largest to occur during the last week of July was a magnitude 3.6 earthquake centered in Perry, Oklahoma.

On the last day of July there were two earthquakes recorded in Oklahoma. The first was a magnitude 2.5 earthquake that was centered in Yale, the second a 3.1 earthquake centered in Blanchard. So despite the Oklahoma Corporation Commission's efforts, Oklahoma is still rocking and rolling.

PART 3
LANDSLIDES

A landslide is a form of mass wasting that includes a wide range of ground movements, such as rock falls, deep failure of slopes, and shallow debris flows. In some parts of the United States, such as in the rolling hills of Southern California, landslides can be a constant threat. Near populated areas, landslides cause an estimated 25 to 50 deaths and $3.5 billion in damage each year in the United States.

Gravity is the primary driving force for any landslide, but there are other contributing factors affecting the stability of the slope. Typically, factors build up specific sub-surface geologic conditions that make the slope prone to failure. The actual landslide just requires something to trigger the slope failure.

Landslides occur when the slope changes from a stable to an unstable condition. A change in the stability of a slope can be caused by a number of factors, acting together or alone.

Landslides have three major causes: geology, morphology, and human activity.

- Geology refers to the characteristics of the subsurface material, the shallow soils and the underlying bedrock. The rock might be weak or fractured, or different layers may have different strengths and stiffness. Sometimes a weak layer of clay can cause a weakness in the rock, forming a plane along which a landslide can occur.

- Morphology refers to the structure of the land. For example, slopes that lose their vegetation to fire or drought are more vulnerable to shallow failure landslides. Vegetation holds soil in place, and without the root systems of trees, bushes, and other plants, the land is more likely to slide away.

- Human activity, such as agriculture and construction, can increase the risk of a landslide. Construction, irrigation, deforestation, excavation, and water leakage are some of the common activities that can help to destabilize a slope. The weight of construction can destabilize a slope, causing a landslide.

Almost every landslide has multiple causes. Slope movement occurs when forces acting down-slope (mainly due to gravity) exceed the strength of the materials that compose the slope. Causes include factors that increase the effects of down-slope forces and factors that contribute to low or reduced strength. Landslides can be initiated in slopes already on the verge of movement by rainfall, snowmelt, stream erosion, changes in the groundwater levels, earthquakes, disturbance by human activities, or any combination of these factors.

Landslides can be classified in many ways. The term landslide is typically used to loosely describe any form of land surface movement down a slope. The rate of movement can range from mere inches a year to extremely rapid. Landslides can be composed of bedrock and unconsolidated or loose sediment or soil. Common landslides can be classified by type.

- Rock Fall: Free fall of material from a vertical slope.

- Rock Slide: Movement of the material is parallel to a zone of weakness and occasionally parallel to the slope. Commonly call a rockslide.

- Slump: A very common type of landslide involving a complex movement of material down a slope. Sometimes called a "rotational slump."

- Creep: The slow movement of material down a typically gentle slope.

Shallow landsides can often occur in areas that have slopes with highly permeable material overlying low permeable material. During periods of high rainfall or lawn watering, water can collect on top of the low permeable material, forming a "perched water table." This results in high pore pressure in the overlying material, reducing cohesive strength. As the upper layer of material fills with water and becomes heavy, the slope can become unstable and slide down slope over the low permeable material.

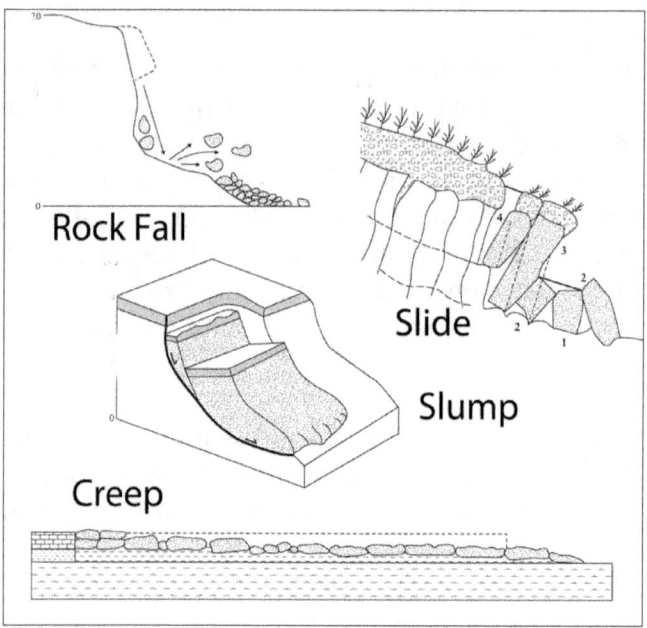

Types of Landslides

The Bluebird Canyon Landslide

Prior to World War II, California was primarily an agricultural economy with a relatively small population. Land was plentiful and cheap. During the war the defense industries brought millions of people to the state to live and work. Many soldiers and sailors passing through California on their way to and from the Pacific war liked what they saw, and decided to stay after the war.

My maternal grandparents moved to California from Iowa to work in the defense industry, as did my father from Illinois. My father met my mother during the war.

During the years immediately following World War II, the "baby boom" years, the construction of homes needed to house the tremendous increase in population quickly used up most of the flat, easily built land. Builders started turning to less suitable land, which usually meant the many hillsides that had previously been mostly ignored. The concept of "homes with a view" quickly became popular, becoming both desirable and expensive.

Landslides are prevalent in southern California because of both the geology and a topography characterized by hills and valleys.

The most common type of landslide is called a slump, because the earth simply "slumps" down the slope. These types of landslides all share the same basic features. The slide typically moves down slope on a zone of weakness in the underlying rock. Where the slide pulls away from the remainder of the hillside above there is a large scar, called a head scarp. Along both sides of the slide are lateral scarps. The surface on the slumped ground itself, the body of the landslide, becomes broken and hummocky. At the downhill end of the slide there is the toe, the mass of earth pushed ahead of the moving ground until it stabilizes by buttressing itself.

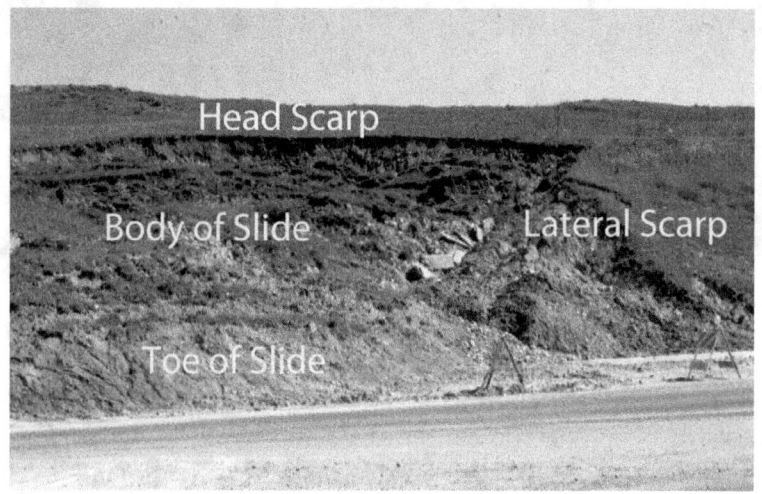

Small Landslide Showing Features
Characteristic of Slumps

By the late 1970's Southern California had been in a drought for more than thirty years. That changed in 1978, and while the rain was needed, too much came too fast. The San Diego River overflowed its banks for the first time in decades, flooding the shopping and office district of Mission Valley. The heavy rains caused slopes overloaded with water to start moving downhill in a number of small and large landslides, carrying some homes with them.

Many of the landslides that occurred in southern California in 1978 actually originated 10,000 or more years ago during the very wet period that was the last ice age. The sliding of the land naturally stabilized the slopes made heavy with water. Over time erosion had altered the classic telltale shape of the typical landslide, hiding its existence from everyone but the trained and experienced eye of the geologist.

Prior to the promulgation of hillside building code regulations in the 1950s, many homes were built on pre-existing landslides that were either not recognized or were ignored. Prior to construction there was typically no geological study of the hillside building site to check for geologic hazards. After the regulations came into effect requiring studies to identify and, if necessary, "neutralize" existing landslide features, many ancient landslides were "stabilized" through "creative engineering" techniques to permit the construction of homes.

Hillside home lots are typically prepared using the "cut and fill" technique. A notch would be cut into the slope to create a flat building site. The soil and bedrock removed from the notch would be piled and compacted on the slope below the notch, expanding the building site. While effective at creating flat building pads on hillsides, the "cut" in "cut and fill" development invariably oversteepens hillsides.

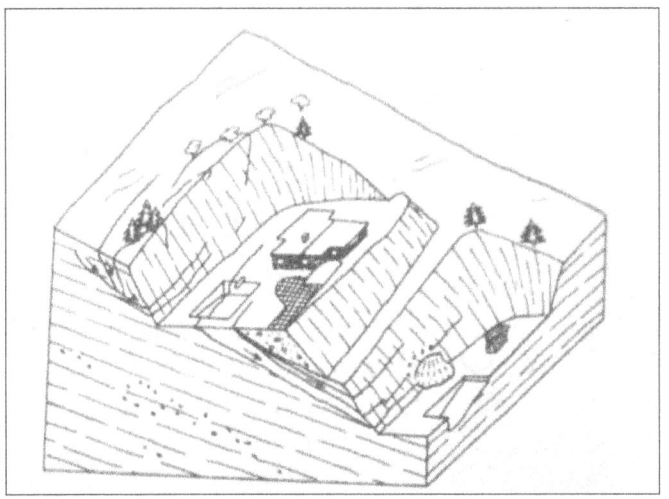

Typical Hillside Cut and Fill Lot Construction

Sometimes, especially on steeper hillsides, cut and fill cannot build a pad larger enough to accommodate a house. In those cases people become more creative, cantilevering their homes over the edge, supporting the house with pilings, like living on stilts.

House Cantilevered Over Pilings

People started building homes on the hillside of Bluebird Canyon starting in 1950, prior to the hillside building codes. Years later, when the geology of the area was being mapped, the slope of the Bluebird Canyon area was recognized for what it was, an ancient landslide. But nobody bothered to tell the homeowners that the ground their homes were built on was part of a landslide. There was a fear that crying "landslide" would depress property values. And, of course, for 25 years homes had been built on the ancient landslide with no harm done.

For many years the construction of homes on hillsides seemed to work. Then it started to rain. Waterlogged slopes all over southern California began to move, reactivating ancient landslides throughout the San Diego and Orange County areas. Many of these landslides were small, but a few were large and involved structures.

This is a Small Landslide

This is a Big Landslide

The reactivation of these ancient landslides happened for three reasons. First, most of the Post-World War II hillside construction had occurred during the "dry" period that started in about 1945. Then when the climate turned wet in 1978 the excess water added weight to the ground and lubricated pre-existing slide planes, the weaknesses in the rock along which the landslide moved.

Second, building construction also added weight to the ground, while other activities such as road construction altered the landscape by removing the natural "buttressing" at the downhill end of the ancient slides, the toes, making them unstable.

Lastly, the construction of cut and fill building sites oversteepened the slope, destabilizing it.

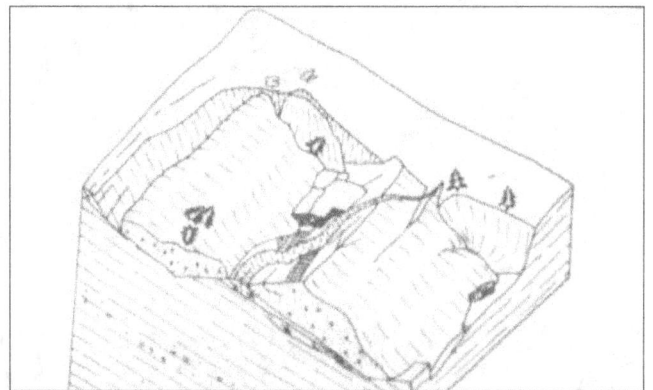

Destabilized Slope Fails in a Slump

Large Landslide in San Diego in 1978

The largest, most destructive, and arguably the most famous of the landslides during this time was the Bluebird Canyon slide. This landslide occurred in a densely populated suburb of Laguna Beach in Orange County, California.

The Bluebird Canyon landslide started early in the morning of October 2, 1978; something went bump in the night to warn residents of an unfolding disaster. During the next few hours, 19 homes and parts of 14 others were carried down slope at an initial rate of 40 feet per hour. Everything moved, the ground, houses, roads and cars. In some places the scarp at the head of the landslide, where the sliding land separated from stable land above, was 35 feet high. Utilities were severed in many places, and city streets were heavily damaged and, in some areas, completely displaced. When the land started moving, tension was put on power lines and the power poles themselves, causing many poles to snap and power lines to fall to the ground. The sounds of power poles breaking were what first alerted many residents to the fact that something was happening.

Damage Along the Lateral Scarp
Note the Water Pipe Pulled Out of the Ground

Power Pole and Lines Pulled Down
By Movement of Landslide

The Bluebird Canyon landslide was called a "block glide" slide because the body of the slide moved as a relatively intact block along a planar rupture surface. The landslide occurred in siltstone and sandstone rocks of the Topanga Formation. In this area the slope of the ground is the same as the southerly dip of the rocks in what is known as a dip slope.

The base of the ancient landslide, the slide plane, was along a zone of weakness in the bedrock, a clayey siltstone at the base of the sandstone that the homes had been built on it. The slope had been made unstable by erosion of the toe of the ancient landslide by Bluebird Creek, removing the buttress and destabilizing the ancient landslide. Water from heavy rains that infiltrated the bedrock underlying the slope lubricated the ancient slide plane. It was then just a matter of time before the slope started sliding down hill.

In 1978 I was working as a geologist with the San Diego County Integrated Planning Office. Because San Diego County was having its own share of landslide problems, shortly after the Bluebird Canyon Landslide I was sent to Laguna Beach on a fact-finding mission. I photographed a lot of the damage that had occurred; these photographs were used in presentations to the public about the hazards that can occur with hillside development.

Displaced Pavement Showing Fallen Utilities

I saw that the worst damage occurred where houses had been built across what became the scarp of the landslide itself. When the landslide occurred, places where the scarp opened under houses caused them to split apart. When the ground beneath a house sitting astride the scarp moved down slope, part remained on stable ground and the other part moved downhill as part of the slide body. Literally the back of the house broke open and the contents simply spilled out.

Aerial View of Bluebird Canyon Landslide Area

House Split Apart by Landslide

On the main body of the landslide the ground itself, centered on Meadowlark Lane, cracked opened in lawns and asphalt streets. Concrete driveways buckled as the ground in the body of the slide shifted. While houses on the main part of the slide itself did not get split in two, the opening of cracks in the moving earth and the formation of hummocky ground heavily damaged them as they shifted off their foundations. Buckled sidewalks and curbs made it impossible to get vehicles out during the evacuation. At the toe of the landslide, the terminal end of the slide, the earth had been plowed up into a high berm.

House Split Open and Contents Spilled Out

When the Back of This House Fell Open the Piano Fell Out
It is in the Lower Right Corner, Sitting Upright

View of the Body of the Landslide from Displaced
Meadowlark Lane at Head of Landslide

By the end of the day the landslide had ultimately affected an area of 3.6 acres, destroying or damaging 50 homes. The total cost of the damage was about $15 million in 1978 dollars. For the people living in the affected area the destruction was total.

Damage to Driveways and Curbs on the Body of the Landslide

View Out the Back of the Garage of One House. Note the Lawnmower in Place on the Left Side of the Photograph

When you buy property, you are buying coordinates in space as shown on a map. If your home moves downhill on a landslide, your house is now sitting on someone else's space. Your property remains where it was in space, now located on a scar left behind by the landslide.

Soil and Pavement Piled Up at the Toe of the Landslide

Edge of the Toe Bulldozed Right Through This House

Because the landslide either heavily damaged or completely displaced the streets, people had to evacuate on foot. They became refugees; all they could take with them was what they could carry. It would be some time before people could get in to retrieve their belongings and automobiles. It was a sad situation all the way around.

The slide that occurred on October 2, 1978 took in just the western part of what had been identified as the ancient landslide. A few months after the slide, on February 9, 1979, a small secondary landslide occurred on a portion of the head scarp.

Written on the Window in Lipstick is a Poignant Message:
"We have taken everyone down."

In 2005 another landslide just east of the 1978 slide, again coming after a period of heavy rain, destroyed or damaged another 19 homes. This landslide took in the eastern part of the same ancient landslide on which the 1978 Bluebird Canyon landslide occurred.

Damage to Homes During the 2005 Laguna Beach Landslide

Damage to House During the 2005 Laguna Beach Landslide
Note the Magnificent Ocean View

PART 4
SEACLIFF RETREAT

Living Right on the Edge

Erosion is a geologic fact of life. It is a natural force that works continually to lower the elevation of the land to sea level. Over millions of years mountains are raised upwards by tectonic forces only to be eroded away by rivers flowing to the sea. While landforms may seem to be perpetual to humans, they are ephemeral in the millions of years of geologic time.

There are, however, places where the effects of erosion are much more noticeable. The coastal bluffs bordering portions of the coast of San Diego are one such example. The near vertical seacliffs that comprise the face of these coastal bluffs are very dynamic, always changing in the battle between land and sea. And this is a contest that the seacliffs will not win.

After a single storm while tide is high, one can see the carnage: chunks of seacliff torn away from the coastal bluffs by the relentless battering of the sea. This is what we call "seacliff erosion," a process that invariably leads to "seacliff retreat." Seacliff retreat has resulted in structures being undermined by the crumbling bluff.

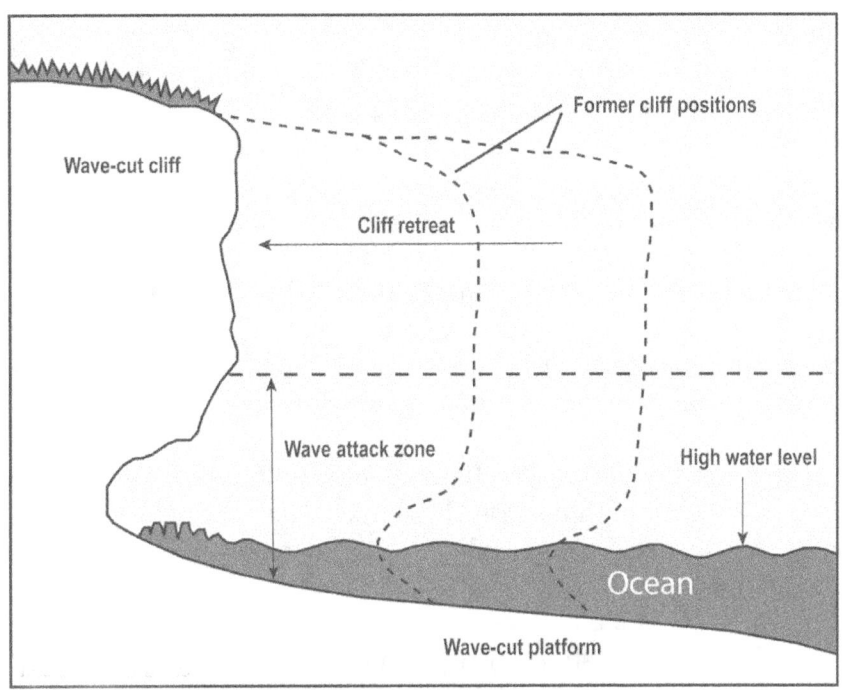

The Retreat of a Coastal Bluff

Development Along the San Diego Coastal Bluff

One such part of the San Diego coast where seacliff retreat is very apparent is an area called Sunset Cliffs. My first visit to the Sunset Cliffs was in the fall of 1973 while attending a class in engineering geology at San Diego State University. A year later I was working as a geologist for the San Diego County Integrated Planning Office (IPO). One of the tasks I was assigned at IPO was to study seacliff retreat over time at Sunset Cliffs. The goal of this work was to develop a County Ordinance determining how close a new structure could be built to the cliff edge and still be there after an estimated building's lifespan of 50 years.

And such an ordinance was definitely needed because people too often wanted to build their homes as close to the cliff edge as they could for the view. In other cases the lots were so small, squeezed between the street and the edge of the sea cliff, that there was no choice but to build close to the edge of the cliff. Erosion of the seacliff at Sunset Cliffs during the 1970s resulted in some rather spectacular failures along portions of the San Diego coastline.

Prior to World War II there was little interest in building along the coast of Southern California. That changed during the years after the war when people developed an appreciation for living near the beach.

Rapid population growth along the coast resulted in a substantial increase in coastal bluff development with little regards to either land use planning or the geology of the building sites.

Living on the Edge, with
Through the Floor Access to the Beach

Until the 1970s there had been limited understanding of the rate at which the seacliffs erode. Nor has there been much thought given to measures to protect both the landowners and the coastal bluffs themselves.

Naturally, people typically wanted to build their homes as close to the cliff edge as they could for the view. Over time the lots shrank in size as the seacliff eroded, sometimes leaving the homes hanging over the edge of the bluff. In other cases empty lots were so small, shrinking in size with time due to seacliff retreat and squeezed as they were between the street and the eroding edge of the coastal bluff, that there was no choice but to build close to the edge of the cliff. And there were plenty of lots for sale, with people paying big bucks for land that may be gone in 20 years.

Seacliff Retreat Leads to Homes Hanging Over the Edge

So erosion of the seacliff has endangered the stability of a lot of expensive real estate. With time, erosion of the seacliff by the pounding of the sea moved the edge of the cliff closer and closer to the homes, sometimes even undermining them, until they were either removed or fell into the ocean. It was a battle of humankind versus the sea. The ocean waves are constantly eroding the seacliff and pushing it farther inland, while people were trying to prevent that from happening.

There are several factors that contribute to seacliff retreat. One of the most important is the geology of the seacliff itself. At Sunset Cliffs the geology plays a critical role in the rate at which the coastal bluff is eroded and the seacliff itself retreats

The coastal bluff at Sunset Cliffs is composed of two geologic formations. The lower part of the cliff face consists of the Point Loma Formation, rock that was originally deposited under the ocean 70 to 75 million years ago. This material is dark in color and is rather resistant to erosion.

Directly overlying the Point Loma Formation is the Bay Point Formation. This loose and easily eroded material is beach sand left behind when sea level was higher 120,000 years ago. The contact, or boundary, between the two formations is flat, representing the sea floor at that time.

The loose sand of the Bay Point Formation is very easily eroded. With just your hand you can scrape out a chunk of loose sand; people sometimes write their names in the soft sand, unthinkingly adding to the erosion.

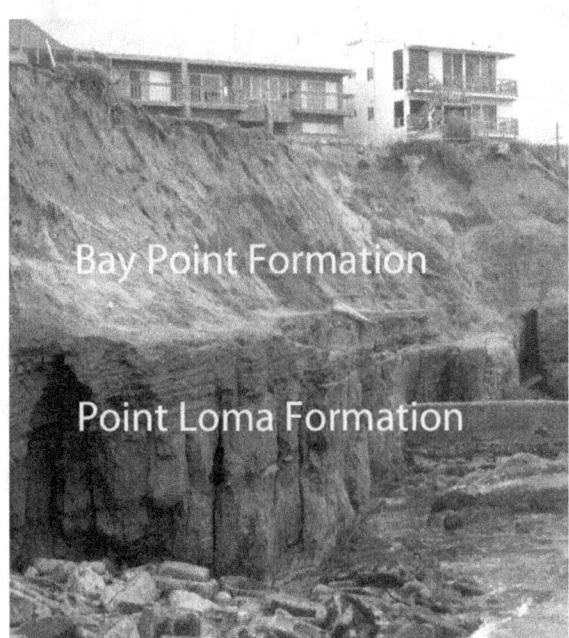

Rock Formations that Make Up the Seacliff at Sunset Cliffs

While the Point Loma Formation is rather resistant to erosion, it is highly fractured by crisscrossing vertical cracks that extend deep into the seacliff. Where the sea has been able to get inside the fracture planes erosion occurs much faster. The pounding of the surf will enlarge the fractures until they become indentations or alcoves in the coastal bluff.

During high tides and storm surges the seawater can work its way into the base along fractures and erode the overlying softer material from below. Erosion can eventually isolate huge blocks of the Point Loma Formation from the seacliff, resulting in the formation of sea stacks.

Another critical factor in determining the rate of seacliff retreat is the climate. During periods with few winter storms in the Pacific Ocean, the Point Loma Formation at the base of the seacliff does a good job of protecting the coastal bluff from erosion. During such times the seacliff can survive the season with nary a scratch. During periods of strong storms in the Pacific, with large swells breaking against the seacliff, erosion can be severe. This is most pronounced during El Nino events, which produce stormy seas. The worst-case scenario is when heavy storm surf coincides with a high tide. During these events the surf can reach up on the seacliff and tear away the soft sands in the Bay Point Formation.

Another factor related to climate is sea level changes over time. Looking at the seacliffs in a longer-term view, the morphology of the San Diego coastline has also changed with the rising and falling of sea level during the last tens of thousands of years. Sea level changes due to the coming and going of ice ages during the last million years have been dramatic. During a warm interglacial period 120,000 years ago sea level stood 20 feet higher than where it is today. At the height of the subsequent ice age sea level was 400 feet lower than today. At that time the coastline was miles farther offshore than at present. When the last ice age ended about 18,000 years ago and the ice sheets and glaciers started melting, sea level slowly rose to where it is today. Sea level is still rising as water locked up in glaciers on the continents and in the polar caps continues to melt.

The coastline we see at San Diego today is the product of the rise and fall of sea level relative to the land. Scientists believe that we are in an "interglacial," a warm period between ice ages. Melting ice has been slowly raising sea level since the last ice age ended. Add the affects of global warming caused by human activity and sea level is rising even faster. Will sea level rise to a point 20 feet above where it is today like it did during the last interglacial 120,000 years ago? Only time will tell.

Yet another factor in the retreat of seacliffs is mechanical erosion, erosion caused by people. People climb up on the more gentle slopes of the Bay Point Formation, knocking loose even more of the sand. In some places, people carve their names in the loose sand, sometimes right below an already undermined building.

When thinking about buying shrinking real estate, remember that when you buy land, you are buying an area on a map located by coordinates. If your land crumbles and falls into the ocean, you still own the same space on the map, only now some of that space is under the sea.

The strength of the material in the seacliff is the predominant factor in determining the rate of seacliff retreat. Average rates of seacliff erosion have been estimated to be little more than one inch per year in the harder Point Loma Formation at the base of the seacliff to nearly 1½ feet per year in the softer Bay Point Formation.

When I was at San Diego County we used an average rate of 0.5 foot per year of seacliff retreat for planning purposes. During a 'quiet" year, when there are few, if any, strong storms lashing the coast, little or no overall erosion may occur. Substantial erosion can occur during a stormy year, especially if coinciding with an El Nino event.

One big storm with high surf, especially at high tide, however, can remove several feet of cliff face. This means that if your house sat 10 feet from the edge of the cliff before the storm, it was five feet away after a big storm like those that occurred in 1978.

One task I was assigned was to help determine how close a structure could be built to the cliff edge and still be there after an estimated building's lifespan of 50 years. Given an estimated average rate of seacliff retreat of 0.5 foot per year, that meant a setback of 25 feet would be required for a structure. There was a lot of resistance against the proposed ordinance because some lots weren't big enough to build a structure with a 25-foot setback; some were mere slivers of land squeezed between the street and the crumbling bluff.

To protect valuable seaside real estate, engineered structures have been constructed along the seacliff during the last several decades. Some of these structures have been successful; most have not.

To protect structures already in place, some people tried to stave off the inevitable by driving pilings into the rock to support their buildings. Other people put cribbing to hold back the lose sand of the Bay Point Formation. Others put large rocks, riprap, at the base of the seacliff to prevent the surf from eroding it. Still others sprayed concrete on the surface of the seacliff to hold it in place. And others built expensive seawalls at the base of the seacliff to prevent erosion. None of these measures have stood up against the relentless pounding of the ocean. In the 1970s there were some examples of failed seawalls at Sunset Cliffs.

As part of the county effort I compared sea-cliff retreat between 1973 and 1979.

One apartment building clearly showed the effects of sea cliff retreat. In December 1973 the edge of the cliff was still several feet from the building, which was still occupied.

In December 1978, five years later, the sea cliff had retreated more than ten feet at this one location, which was far more than the average rate of retreat. The foundation of the building had been undermined and the building was abandoned. This apartment building had become a derelict, a hazard to beach goers. A year later the building was gone.

Undermined Houses Supported on Pilings

Example of a Failed Concrete Covering on Seacliff

Remnant of Failed Seawall at Sunset Cliffs
Note the Name Scrawled into the Loose Sand

View Along Sunset Cliffs Showing Fractured
Point Loma Formation and Sea Stacks.
Note Failed Seawall Lower Left of Center

Apartment Building in 1973

The Same Apartment Building in 1978

Profile View of Apartment Building in 1978

*The Main Part of the Apartment Building is
Gone in Early 1979*

When I gave a presentation to a citizen group in early 1979 one person in the audience was the owner of the apartment building that had been removed. He said that it had been a difficult thing to watch the edge of the seacliff move closer to the building with every storm, then to consume it. It was a total loss.

Another man owned an empty lot, or what was left of a lot, that he wanted to build on. He was against the proposed 25-foot setback ordinance because there would be no room for the house he wanted to build. He proposed setting the foundation on deep pilings so that when erosion had consumed the lot the house would still be standing, on poles. If he wasn't allowed to build that way the lot would be good for nothing more than a small picnic ground. The house was never built.

In the end it made no difference. In 1983 the stretch of Sunset Cliffs I had studied was turned into the Sunset Cliffs Natural Park. All of the buildings that had been clinging to the edge of the cliff were removed.

Sunset Cliffs in the 2000s.
Note Riprap Piled Against Seacliff to
Prevent Erosion

More than forty years after I first visited Sunset Cliffs in 1973 the local authorities are still trying to stabilize the seacliff. Every year the 68-acre Sunset Cliffs Natural Park gets a little smaller as the ocean nibbles at the seacliff.

PART 5
ON THE FRONT LINES OF GEOLOGY

My professional career as a geologist began in September 1974 when I got a job with the County of San Diego's Integrated Planning Office. I had graduated from San Diego State University with my Bachelors Degree in Geology the previous June, and in September was working on my Masters Degree.

The Integrated Planning Office was working on the implementation of the California Environmental Quality Act (CEQA) in San Diego County. My assignment was to study both geology and groundwater availability in sub-regional community planning areas and to make recommendations on permissible population densities given the groundwater resources available.

I spent a lot of time in the rural areas of San Diego County studying the geology and measuring water levels in water-supply wells. I eventually wrote reports on groundwater availability in the Jamul, Ramona, and Julian Subregional Growth Management Areas. I also made presentations on groundwater to community groups, which was a lot of fun. This was the start of my career, although being on the regulatory side of the business was more about "observing" geology rather than "doing" geology.

It was while working for the Integrated Planning Office that in 1975 I first met Kathy, the woman who would become my wife two years later.

Later in my career at San Diego County I transferred to the Environmental Analysis Division where I reviewed environmental impact statements for various citizen and public projects. In 1978 I transferred to the Regional Growth Management Office to work on the hydrogeological parts of the San Diego County growth management plan. When time permitted I also assisted the public in determining the locations for water supply wells.

After I got my Masters Degree in Geology in 1977 I also got a job teaching evening adult education classes in geology at San Diego Mesa Community College. I taught from September 1977 through June 1979. I really enjoyed teaching and the academic environment.

My wife Kathy and I moved to Oklahoma City from San Diego in 1979 so I could accept a position as hydrogeologist with Kerr-McGee Corporation, an energy company based in Oklahoma City. I accepted the position because it was a good opportunity for me professionally and was a good move for Kathy and I.

During my time at Kerr-McGee I worked on the environmental side of the business. To corporate management those of us doing environmental work were a "necessary evil" mandated by state and federal regulations, but costing the bottom line.

At Kerr-McGee I was involved in a lot of projects where I was assessing the extent of environmental contamination caused by past and present company operations. I also designed and supervised the construction of water wells to pump contaminated groundwater to the surface for treatment. I gained notoriety as a person who always brought the project to a successful conclusion.

If someone else was having problems getting their project done, I was often sent in as a "trouble shooter" to take the project over. I would get obsessed with getting the job done, and had a reputation for doing just that. I gained the respect of management for this, and made enemies among people I displaced to get a project done.

While at Kerr-McGee I got to see and "do" a lot of interesting geology, although being on the environmental side of the business a lot of the geology I saw was what I would call "dirty geology." That is, geology contaminated by any number of different "bad actors."

I also got to see a lot of the country I'd never seen before. I traveled to Wyoming a lot and fell in love with the state. I also did work in Arizona, Florida, Illinois, Kansas, Louisiana, Montana, Nevada, New Mexico, Mississippi, Oregon, Tennessee, Texas, and Wisconsin as well as in Oklahoma. I worked in deserts and plains and mountains and cities. I worked in nice weather, hot weather, freezing temperatures, rainstorms and snowstorms.

Our Hydrology Department had a very tight knit group of experienced and committed professionals who enjoyed working with one another and who helped each other whenever necessary.

Kathy liked my involvement in environmental issues, which I was actively engaged in when we met. People have asked me, often nastily, how I could work for an oil company; an industry known best for making environmental messes. After all, the environmental movement started in earnest in 1969 after the infamous Santa Barbara offshore oil platform blowout.

My reply to my detractors is always - what's the point of belonging to an organization of environmentally conscious people who hang out with one another and complain about the situation, but never try to do anything to change it other than protesting with a sign in front of an industrial facility gate? I spent nearly five years in California working with people like that. I spent a total of 22 years in Oklahoma working for environmentally unaware people assessing and cleaning up environmental messes, actually doing something productive, something I am proud of. I think I did make a difference during my professional career.

My primary responsibility during most of my time at Kerr-Mc Gee was doing work for the Kerr-McGee Coal Corporation. Kerr-McGee had three coalmines in northeastern Wyoming near Gillette, only one of which was operating at the time. Most of my work was for the Jacobs Ranch Mine located south of Gillette, which in the 1990s was the second largest open pit coalmine in the United States. I had a very close and good working relationship with the Coal Corporation people, who respected me and appreciated what I did for them. And I enjoyed working with people in field offices because they were casual (no suits) and low-key compared to the uptight (Kathy called it "anal retentive") atmosphere in the Oklahoma City corporate office.

Kerr-McGee's coalmines were located in the Powder River Basin in northeastern Wyoming. The Powder River Basin is a topographic drainage and geologic structural basin that extends from southeast Montana through northeast Wyoming. The basin is so named because it is drained by the Powder River, although it is also drained in part by the Cheyenne River, Tongue River, Bighorn River, Little Missouri River, Platte River, and their tributaries.

The Powder River Basin is known for its coal deposits. It is the single largest source of coal mined in the United States, and contains one of the largest deposits of coal in the world.

Wyoming has been the top coal-producing state in the United States since 1988. The United States uses about 1 billion tons of coal a year, supplying about a third of the United States' electricity demand. The coalmines in the Powder River Basin annually mine more than 400 million tons of coal a year, about 40 percent of the total United States usage, more than twice the production of second-place West Virginia, and more than the entire Appalachian region.

The coal in the Powder River Basin is classified as "sub-bituminous," having an average energy potential of approximately 8,500 btu/lb, with low sulfur dioxide. Contrast this with eastern, Appalachian bituminous coal containing an average of 12,500 btu/lb and high sulfur dioxide. Even with its lower energy potential, Powder River Basin coal became valuable in the 1970s when air pollution emissions standards for power plants became a concern. Because of its lower sulfur dioxide content, Powder River Basin coal used in coal-fired power plants burns much cleaner than does Appalachian coal.

The majority of the coal mined in the Powder River Basin is part of the Fort Union Formation of Paleocene age. The coal beds began to form about 60 million years ago when the land began rising from a shallow sea. When the coal beds were forming, the climate in the area was subtropical, averaging about 120 inches of rainfall a year.

For some 25 million years, the basin floor was covered with lakes and swamps. Because of the large area of the swamps, the organic material accumulated into peat bogs instead of being washed to the sea. Eventually the climate became drier and cooler. The basin filled with sediment that buried the peat under thousands of feet, compressing the layers of peat and forming coal. Over the last several million years, much of the overlying sediment has eroded away, leaving the coal seams near the surface.

While working for the Kerr-McGee Coal Corporation I designed and supervised the construction of water-supply wells at the surface coalmines. These wells, which went as deep as 2,200 feet, produced water for dust control in compliance with state and federal regulations, as well as for coal processing and drinking water uses.

In addition to installing water-supply wells deep in the Fort Union Formation, I also prepared coalmine geology and hydrogeology reports and special reports addressing specific issues regarding the underground water-bearing formations affected by surface coal mining. I also wrote the hydrogeology chapters of the state mine permit documents, which included computer models predicting the impacts of mining on groundwater levels; the state of Wyoming mandated that mining was not supposed to lower water levels in local rancher's water-supply wells.

As part of the coalmine permitting process I did analyses of climate, rainfall and snowmelt runoff, groundwater replenishment and storage characteristics, and the testing of water-bearing formations. And I designed and supervised the expansion and improvement of the system of groundwater monitoring wells designed to show if water levels were being affected by mining in compliance with state regulations.

Over time I gained a reputation for both the success and accuracy of my work as well as my efforts to make sure that the state government was fair and accurate in their regulating of mining. I received a number of accolades from the Coal Corporation for this work.

Steve Doing Pump Test at the
Jacobs Ranch Mine in 1981

When Kerr-McGee sold the Coal Corporation unit in 1998 I was hoping that I might be offered the opportunity to leave Kerr-McGee and move to Wyoming to work for the new owners of the Jacobs Ranch Mine. Kathy would have absolutely loved living in Wyoming; she had been born in Laramie, Wyoming in 1947. I never received an offer to make the move, though, which disappointed me. It wasn't until a couple of years later that I learned that the new owners of the Jacobs Ranch Mine had indeed asked Kerr-McGee corporate management to let me go so I could continue to do the work I'd been doing for them for 19 years, but that Kerr-McGee management had refused to let me go. Nobody had bothered to ask for my opinion.

When not doing projects for the Coal Corporation I did projects at other Kerr-McGee facilities. Projects I was involved in ranged from installing water-supply wells at remediation sites where contamination such as gasoline was being removed from groundwater to irrigation and fire control purposes at operating Kerr-McGee facilities. Whenever a Kerr-McGee operation contaminated a person's water supply well I made sure they got a new well and a clean water supply.

TEAMING UP

Corporate and coal staff defeat flawed groundwater study

Steve Lower travels to Jacobs Ranch Mine a couple of times a year and occasionally visits Galatia. His visits go unnoticed by most employees. However, the environmental staffs are quick to welcome him. Monitoring wells, bore holes, water quality problems and permits are synonymous with Lower, as he is the corporate groundwater expert consultant for K-M Coal projects.

Lower, a staff hydrologist in the Safety and Environmental Affairs Division, helps prepare permit applications, annual groundwater monitoring reports and special groundwater reports that both mines must submit to regulatory agencies. Lower works closely with Clem Burdick, director of environmental affairs; John Coleman, manager of environmental affairs; and the environmental staff at the mines to ensure environmental compliance. Darryl Maunder and Roy Liedtke at Jacobs Ranch and Gerald DeNeal at Galatia rely on Lower for prompt, reliable groundwater problem-solving.

Lower's environmental expertise recently played a key role in the defeat of a regulatory groundwater pilot study. In 1993, federal and Wyoming regulatory agencies initiated development of a pilot computer model to predict groundwater drawdown impacts from mining activities at Jacobs Ranch and two neighboring mines. The results over-predicted impacts compared with present observed conditions, indicating that future predicted impacts would also be excessive. Lower's review and critique of the study cited inaccuracies that explained why the current and future drawdown projections were incorrect.

Coal's his cause: Steve Lower works on another Coal environmental report.

Following a hearing at which Lower reiterated his concerns and recommendations, the Wyoming Department of Environmental Quality agreed with Lower and abandoned the model for mine permitting use.

"Steve's been a valuable asset to K-M Coal," Burdick said. "The defeat of the pilot study was a major victory for K-M Coal as well as for the Wyoming coal industry. This model was a real threat because it incorrectly over-predicted groundwater impacts, which could have delayed permits and interfered with our mining."

Lower's enthusiasm for coal is evident. He currently serves as chairman of the Gillette Area Groundwater Monitoring Association (GAGMO), which coordinates coal industry data collection and provides technical oversight for groundwater regulatory developments. "Steve has been a part of GAGMO since its inception, and his continuing work with the organization has given Kerr-McGee an invaluable leadership role in the Wyoming groundwater regulatory area," said Dick Turpin, engineering manager at Jacobs Ranch.

"I've been working with the Coal Corporation on various projects since 1979," Lower said. "I enjoy working with the people in Coal, and I'm proud to know that I've been able to help them out."

In 1984 I started Kerr-McGee's first efforts to clean up gas stations where gasoline had leaked from underground storage tanks and piping into the soil and contaminated the groundwater. This gasoline sometimes got into sewer lines and storm drains where fumes could travel great distances, sometimes even into homes. When gasoline got in contact with buried phone lines it would dissolve the wire insulation and cause shorts circuits.

This effort to clean up gas stations was mandated by new regulations promulgated by the U.S. Environmental Protection Agency (EPA). Prior to the new EPA rules no one was too concerned if gasoline was lost through leaks. Sometimes the leaks were in the gasoline storage tanks themselves and, more often, in the piping between the tanks and the gas pumps. The new rules eventually required borings to be drilled and monitor wells installed at all gas stations to determine if gasoline was present in the subsurface soil and rock and the groundwater. Most of the older gas stations I did work at had at least some gasoline in the subsurface.

Cleaning up gasoline spills involved the construction of interception systems designed to capture the gasoline and contaminated groundwater. Where the water table was deep, wells would be constructed to pump out free gasoline and contaminated groundwater. In areas where the water table was shallow, trenches were constructed to remove the contamination.

A gas station I worked at in Tulsa, Oklahoma in 1986 had both shallow gasoline contamination from leaking pipes and deeper contamination from leaking underground tanks. I used both wells and trenches to clean up that gas station over a period of months. Gasoline leaked from this station had made its way into the telephone company's lines, making it necessary to replace all the wires.

In 1984 I performed the first remediation work at Kerr-McGee's former Deep Rock oil refinery in Cushing, Oklahoma. The Cushing area once had been a center of extensive petroleum production and refining; petroleum production, storage; and pipeline and petroleum storage activities continue in the general area.

The Kerr-McGee Cushing remediation site is a former petroleum refinery encompassing approximately 440 acres located about 2 miles north of Cushing. Deep Rock Road divides the site into north and south areas.

Several different owners had operated a refinery on the property starting about 1915. Kerr-McGee purchased the refinery in 1956 and operated it until 1972, after which time they dismantled the refining facilities and most of the storage tanks.

I first began remediation activities at the site with a geomagnetic survey in 1984. This survey was conducted to locate buried pipelines and refinery waste in the subsurface. Some pipelines were found in a tangled weave not unlike spaghetti, many of which were not on the site map. During operation of the refinery it was cheaper and faster to lay a new pipeline than to find and repair a leak in an old line. There was no consideration for the petroleum that had leaked into the ground.

Other than showing the pipelines, reviews of the geomagnetic survey data were ambiguous. This was later found to be because petroleum waste was everywhere we looked.

During those first visits to the Kerr-McGee site in 1984 I found the place to be a mess. Petroleum waste could be found on the surface almost everywhere. Pools of petroleum sludge were found anywhere a storage tank had once stood. There were patches of old, desiccated petroleum waste found along creek bottoms as well as sheens from fresh hydrocarbon seeps glistening on the water. And there were five large pits full of acidic refinery waste.

I would be involved with work at this site on and off from 1984 until my retirement from Kerr-McGee in 2001.

During that time my colleagues and I drilled hundreds of soil and bedrock borings to gain an understanding of the geology underlying the site. Of these borings, a total of about 98 were eventually completed as groundwater monitoring wells to investigate the hydrogeology of the site.

Petroleum Sludge Pool at Former Tank Site

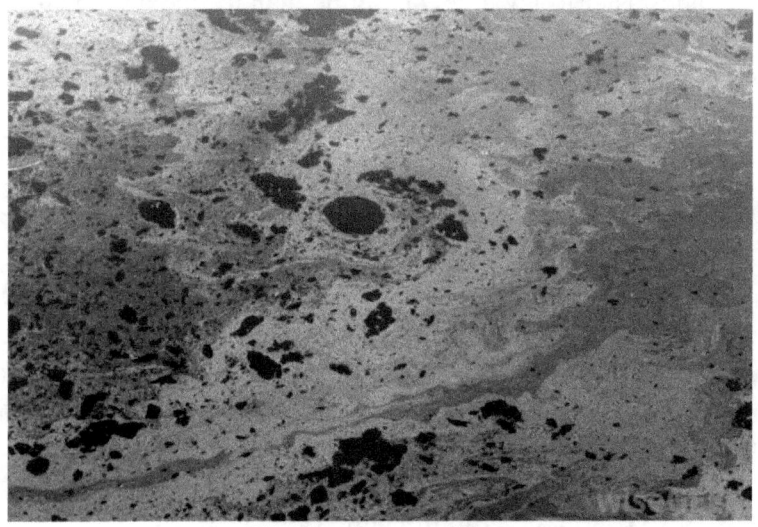

Oil Floating on Water

During my review of topographic maps of the site I found that the remediation site lies entirely within the drainage basin of Skull Creek, a tributary to the Cimarron River. Skull Creek originates about 1½ miles southwest of the Site, cuts diagonally through the site from south to east, and joins the Cimarron River about four miles northeast.

Skull Creek was so named for a reason. Prior to site remediation in the late 1990s Skull Creek was a "dead stream;"

nothing lived in Skull Creek because of contamination by refinery waste. Over the many decades that the Kerr-McGee refinery site had operated, as well as other refineries that were located upstream from the site, the water of Skull Creek was a handy place to dispose of refinery waste.

During my preparation of reports on the geology and hydrology of the site I found that it could be described as a series of highlands overlooking a broad stream valley occupied by Skull Creek. Resistant sandstones and limestones underlie the highlands to the south, while the valley floor is cut through soft stream terrace deposits and underlying soft mudstones to the north.

Research showed that the remediation site is directly underlain by bedrock of Pennsylvanian and Permian age, the last two periods of the Paleozoic Era, ranging in age from 280 to 310 million years ago. What is now North America was then located on the equator; the climate in North America at that time is believed to have been arid or semi-arid.

Investigations of the geology showed that the bedrock immediately underlying the site consists of mudstones, lenticular sandstones, and thin limestones of the Upper Pennsylvanian Vanoss Group. The Vanoss Group is in turn underlain at depth by the sandstones of the Ada Group and Vamoosa Formation, collectively known as the Vamoosa-Ada Aquifer, also of Upper Pennsylvanian age.

The lithology of this bedrock represents cyclic sedimentation during multiple transgressions and regressions of shallow seas across a relatively flat continental interior. These recurring depositional environments resulted in a repetitious sequence of mudstones, lenticular sandstones, and thin limestones deposited during alternating deltaic, tidal flat and shallow marine environments. Due to the relatively flat terrain, small changes in sea level caused shorelines to transgress and regress over distances measured in hundreds of miles.

In those areas between distributary channels on the delta where little sand was being deposited, tidal flat conditions consisting of extensive mud flats were predominant. This type of deposition is thought to be the origin of the mudstones underlying the site. The red color of some mudstones may be indicative of oxidation on mud flats or in shallow water under arid or semiarid climatic conditions. The gray mudstones would have been deposited in deeper water on the delta. Thin limestones were deposited during transgressions of the warm, shallow seas across the mud flats.

The thick sequence of non-fossiliferous red mudstone and overlying non-fossiliferous gray mudstone exposed in the bottom of Skull Creek probably originated on the late Pennsylvanian mud flats. The fact that the gray mudstone becomes fossiliferous and shaley in the upper few feet indicates transition to a shallow marine environment where clay and silt were still being deposited. The presence of the overlying fossiliferous limestone suggests the transition to a deeper marine environment where clay deposition was absent and a limey bottom supported a rich assemblage of oysters, crinoids, and brachiopods.

I began the hydrogeologic investigation of the site with the installation of the first 38 groundwater monitoring wells in 1985. Studies of groundwater circulation gleaned from water level data show that the site is located in an area of groundwater discharge. Skull Creek represents the local base level, and is thus the receiving stream for discharging groundwater, some bearing refinery contamination, as well as surface runoff. Groundwater recharged to the shallow bedrock underlying the highlands surrounding the site is moving down gradient through the site to discharge to the creek.

During my investigations of the Kerr-McGee remediation site, I found refinery waste and product contamination in the soil and underlying bedrock as well as in the groundwater over much of the site. I found contamination in almost every boring I drilled and monitor well I installed. Some borings had free gasoline resulting from leaking storage tanks that had once stood on the site. Other borings had oil in them accumulated from years of refinery operation. Prior to federal regulations on environmental protection, no one was concerned if refinery waste was discharged to the ground and pipes and storage tanks leaked. That resulted in decades of refinery waste and petroleum products such as gasoline seeping into the ground.

Steve at the Cushing
Remediation Site in 1986

The biggest problem, however, involved the five large pits full of acidic petroleum waste.

Prior to 1951, a sulfuric acid treating unit was used to process lubricating oil. That unit produced waste diatomaceous earth and clay filtering materials containing sulfuric acid and petroleum waste. Those wastes were deposited in five unlined pits (Pits1 through 5) on the property. Pits 1 through 4 were located north of Deep Rock Road, while Pit 5 was located south of the road, near Skull Creek. The waste volume was estimated to be 300,000 cubic yards, half of which was contained in Pit 5. After 1951, a different treatment unit was installed, and production of acid wastes stopped.

Pits 1, 2, 3, and 5 were full when Kerr-McGee purchased the refinery in 1956, and, therefore, were not used in its refinery operations. Pit 4 was partly full of acid petroleum waste; Kerr-McGee used Pit 4 to dispose of desalter refinery waste and waxy residues. In addition to the five large pits, in 1994 Kerr-McGee evaluated 21 additional small pits containing petroleum sludge, some of which were former storage tank sites.

The five large pits were originally constructed by rerouting streams. The stream bottom sediment was removed down to the bedrock, then the old channels were dammed. The newly created pits were then filled with acid sludge waste. Rerouting Skull Creek and damming the channel thus created pit 5, the largest of the five pits. Smaller stream channels were dammed to create Pits 1 through 4. Since the pits were not lined, over time sulfuric acid and petroleum waste entrained in the waste seeped into the bedrock beneath the pits. In addition, sulfuric acid and petroleum waste seeped into the newly rerouted stream channels.

Most of the site investigations that have been performed since 1984 have been done to determine the extent of surface and subsurface contamination. Those investigations have focused on groundwater and geologic conditions around the five acid sludge disposal pits, small stream channels associated with the pits, and Skull Creek. Samples were obtained of waste acidic petroleum sludges and tars, soils, sediments, groundwater, and surface water. These activities include rerouting Skull Creek away from Pit 5 and installing underground collection systems to intercept and remove petroleum and acid seepage.

In 1992 I installed a large "French drain" system next to Skull Creek to intercept oil that was seeping into the creek. This system was very effective at collecting oil that would otherwise have seeped into the creek. A second "French drain" system was subsequently installed between the west side of Pit 5 and Skull Creek to intercept acidic water that was seeping into the creek.

In 1985 I also performed the first major investigations to measure soil and groundwater contamination resulting from past nuclear processing at Kerr-McGee's Cimarron plant near Crescent, Oklahoma. This was the site where Karen Silkwood had worked in the early 1970s. The purpose of this work was to determine where the soil and groundwater were contaminated with uranium and plutonium.

In 1986 I performed the first investigation of impacts from past refinery operations at Kerr-McGee's refinery site in Cleveland, Oklahoma. Oil that had leaked into the ground had found its way into a stream that flowed into a nearby reservoir, Keystone Lake. The Army Corps of Engineers had followed the hydrocarbon sheen on the water upstream from the lake to the refinery site. In addition to these first investigations and remedial work to stop the flow oil into the creek, in the late 1990s I supervised the plugging and abandonment of a deep injection well at the Cleveland site that had been used to dispose of hazardous chemicals.

In 1991 I supervised the installation of a water-supply well to fill and maintain the level in a lake Kerr-McGee constructed at their Technical Research Center in north Oklahoma City. In the late 1990s I performed a series of tests on a well to determine the feasibility of removing contamination from groundwater near Kerr-McGee's rocket fuel plant in Henderson, Nevada. In the late 1990s I constructed and tested a water supply well at Kerr-McGee's remediation site in Calhoun, Louisiana. This well was used to help remove contaminated groundwater from the site.

Also in the late 1990s I constructed a water supply well at a Kerr-McGee facility in Wisconsin. This well was used for industrial uses and as standby for fire fighting purposes.

And in 1998 I installed a dewatering system at Kerr-McGee's former nuclear processing plant in West Chicago, Illinois. This system removed groundwater so that soil contaminated with the radioactive element thorium could be removed.

But after Kerr-McGee sold the Coal Corporation in 1998 much of my work was at the Cushing remediation site. One of the biggest projects involved Pit 5, the largest and deepest of the five acid sludge pits. When Kerr-McGee started removing the sludge the pit filled with strongly acidic water; it turned into an acid lake. The contractor brought in a barge from which to drill to the bottom of the pit and determine how much acid sludge remained. The contractor was unable to do this with their equipment, so I took over the project and brought in different equipment. We were then able to drill to the bottom of the pit and determine how much waste remained in Pit 5. Kerr-McGee was then able to complete neutralization and removal of the acid sludge waste.

INTERNAL CORRESPONDENCE

KM-814

TO	R. A. Thompson	**DATE**	August 14, 1998
FROM	T. Gibson	**SUBJECT**	Recognition – Steve Lower

CHEMICAL
West Chicago Project

I would like approval to present a Level 2 – Quality Achievement award to Steve Lower for his efforts in installing the De-watering Wells in Pond 1. Steve's approach to the project was very professional. He arrived with a schedule in mind and a determination to meet it. It can be hard for someone at West Chicago for the first time to find his way through the system, but Mr. Lower managed to do it with minimal assistance.

The work area was very congested, but even when working with an inexperienced subcontractor Steve managed to maintain a safe working environment and cooperated with other team members so that all contractors continued their projects. He continuously communicated his needs and concerns to other team members. He kept the work moving, even when experiencing problems with free flowing sand and uncooperative electrical controls. Steve managed the development of the wells and returned to personally test their performance. He wrapped up his efforts providing a punch list inspection and system status report, together with hands on instructions for the operators.

The entire project seemed to go smoothly and according to plan. Mr. Lower's performance exceeded my expectations in that he was able to do so well with so little assistance. If you approve, I propose to present Steve with a Level 2 award. You will need to advise Steve's supervisor and the Recognition Review Committee. Attached is congratulatory memo from Jeff Williams.

Please approve this Level 2 award.

TG/TG

In addition, I supervised the characterization of acid sludge waste at Pit 4 at the Cushing remediation site. I used data from the borings to estimate how much waste needed to remove from the pit.

Kathy was very proud of the work I did at Kerr-McGee. She kept a file of newsletter articles about me as well as the many letters of praise and accolades and recognition awards I had received from management over the years.

From:	Lux, Jeff
Sent:	Thursday, May 13, 1999 11:10 AM
To:	Holmberg, Harold; Larsen, Jess
Cc:	Widmann, Roy; Nelson, Stephen; Pounds, Robert; Lower, Steve
Subject:	Steve turned it around!

Harold and Jess, I wanted you to know that Steve Lower has pulled a couple of irons out of the fire for us. We recently determined we had to characterize the north end of Pit 5 from a barge, and VFL's QC people turned out not to have the drive to get the job done or the expertise to generate quality data. We asked Steve to step in and take the project over. In modifying the process to make it work better, he reduced the number of people on the barge from six to three, increased the number of holes being drilled each day, and is generating better data than VFL had been generating. Over the course of only a few weeks, Steve will save us thousands of dollars while generating the quality of data we need.

Also, we are investigating an area south of Pit 5 for a disposal cell site. We asked Steve to run the drilling program, and when he got pulled off this work for barge characterization, he brought B&McD in on the project, and we are now making good progress on both jobs simultaneously. Steve is doing a good job keeping both projects going.

I know this degree of involvement causes problems with Steve's availability for other projects, but I feel it is worth it to have an experienced and aggressive individual who is willing to do what it takes to get the right results. In the past, we have too often sacrificed information or quality for expediency, and Steve has really helped us minimize costs while obtaining good data.

GOOD JOB, STEVE!

Steve in the Bottom of Remediated Pit 5
Cushing Remediation Site
August 2000

Perhaps it's typical of corporate America (Kerr-McGee was my only experience with "the beast"), but the atmosphere at Kerr-McGee's headquarters was always uptight at best, more often tense. Or, as Kathy once told me, "anal retentive." Kathy came downtown to visit me a few times to have lunch, but absolutely hated the headquarters offices, and never felt comfortable there. She used to say that she could feel the tension in the air as soon as she entered the building. It's no wonder I preferred working at a field office like the Jacobs Ranch Mine or Cushing where the atmosphere was low-key and people left you alone to do your work. People at field offices were more friendly, and interested in getting the work done. And the field offices appreciated the work I did for them.

Steve in His Oklahoma City Office in 1985
Wearing the Dreaded Suit and Tie

I fell into a deep depression after Kathy's death, and was in no mood to continue my career. Even now, 15 years after her death, I am still in therapy. I had an interesting career, but it is over.

PART 6
MAKING POTTERY IS A GEOLOGIC ENDEAVOR

The Vietnam War interrupted my college education in early 1966. After serving in the Army in Southeast Asia, I enrolled at Fullerton Junior College again in 1971 as a start on my way to finishing my college education. My major in 1971 was still geology, but by this time I was also very interested in the environmental and computer sciences. Classes in environmental science were few in the early 1970's, but I took whatever courses were available in addition to my geology classes. There was one class in computer science, using a primitive computer (remember this was in 1971!). At this same time I also got interested in making pottery.

Making pottery is a geologic endeavor. Geologically speaking, clay is a fine-grained natural material that combines one or more clay minerals with traces of metal oxides and organic matter. The clay forms from the weathering of rocks at the surface.

The Earth's crust is made up largely of igneous rocks, such as the granite that forms the crust and rocks expelled by volcanoes. The Pala pegmatite described earlier in Part 1 is an igneous rock. Clay minerals typically form over long periods of time as a result of the gradual chemical weathering of these igneous rocks by low concentrations of carbonic acid and other dilute solvents. These solvents, usually acidic, migrate through the weathered rock.

There are two types of clay deposits: primary and secondary. Primary clays form as residual deposits resulting from in-situ weathering of rock and remain at the site of formation, such as the clay in the Pala pegmatites in which the tourmaline crystals are found. Secondary clays are clays that have been transported from their original location by water erosion and deposited in a new sedimentary deposit. Such clay deposits are typically associated with very low energy depositional environments such as large lakes, river deltas and deep marine basins. If buried under significant successive layers of sediments, with time clay can be compressed into the sedimentary rock shale.

Clays exhibit plasticity when mixed with water in certain proportions. When wet, clay can be formed into any variety of shapes. When dry, clay becomes firm, and when fired in a kiln, permanent physical and chemical changes occur. These changes convert the clay into a ceramic material. Because of these properties, clay is used for making both utilitarian products such as bricks and wall and floor tiles and decorative products such as pottery. Different types of clay, when used with different minerals and firing conditions, are used to produce earthenware, stoneware, and porcelain pottery.

In the early 1970s I went on a class field trip to the Hopi Indian reservation in Arizona to watch how the Indians made pottery. In the early 1970s the Hopi still made their pottery as it had been done for generations, without using a potter's wheel to make a "thrown" pot. After collecting clay from a natural source on the reservation, the Hopi would roll a piece of the clay between their hands into a length of rope-shaped clay several inches long. The Hopi would then start coiling the clay rope round and round to build up the pot. Once the basic shape of the pot was done the Hopi would flatten the coils on the inside, filling in the gaps between them in the process, resulting in a smooth, polished interior. Sometimes they would smooth over the coils on the outside of the pot, and other times leave the coils showing.

Once the finished pot had a chance to dry, the Hopi would paint it using colors made from naturally occurring minerals found around their homes. After painting, the pot would be buried in burning cow dung in an earthen oven or "kiln" to be baked, a process called "firing." This process heats the soft clay until it becomes hard and ready to use.

After watching this process and seeing the beautiful results, I knew that I wanted to make pottery. Despite the fact that I had never made pottery before in my life, I was going to make a pot using methods similar to what the Hopi were using.

It took me a while to get my project going. Not having a source of natural clay in suburban Fullerton, I bought a block of pottery clay from a craft shop. After many failed attempts I finally got the hang of rolling a piece of clay into a rope without it cracking because it was getting dry; adding water during the process to keep the clay moist was the key. After coiling the clay ropes into a pot shape I learned that wetting my fingers while smoothing the coils on the inside of the pot to give it a polished look worked best.

But there was a tradeoff. I learned early on that if the clay was too wet, or the rope coils too thin, the sides of the pot would collapse. If the clay dried too fast, the pot would crack. Back to the drawing board. It took me a while, but after several dismal failures I had a finished pot. I could take some consolation in the fact that the Hopi pot makers said that a 30% failure rate was common for them.

After the clay had dried, I painted a primitive design on it using commercial iron oxide pottery paint. Being fresh out of dried cow dung in Fullerton, I took the finished pot to Fullerton Junior College where my geology professor, Walter Reiss, who had been making pottery for years, taught me to fire the pot in his pottery kiln. My project was done, although it certainly didn't have near the professional look that the Hopi pottery did. But it was a pot I had made.

I decided to make pottery a hobby. I experimented with many different designs, mostly making the pots using coils of clay, and a few using slabs of clay. I used commercially available glazes, firing the pots in Mr. Reiss' kiln with his blessings. I soon learned that smoothing some coils while leaving others exposed added an interesting design element to the pots. And, I found that giving away pots was a great way make friends with women.

Using Pottery to Teach Mineralogy

As a returning veteran, when I was enrolled at Fullerton Junior College in 1971 I got back the job I had during my first years at the school in the mid 1960s, that of a part-time technician in the geology Department's lab and preparation room. In that position I assisted the geology teachers in the preparation of materials for the geology laboratory classes. I also went on all of the field trips as a teacher's assistant.

In the days before computer operated spectrographic equipment, the borax bead test was used in mineralogy to identify the predominant element in a mineral sample. In the mineralogy lab class, students would ground up a piece of a mineral in a mortar and pestle. A small quantity of the resulting powder would be mixed with borax and shaped into a bead at the end of a rod. The bead was then held over the flame of a Bunsen burner. The color that flared off the borax bead was then compared to a reference chart to identify the element, which could then be used to name the mineral.

In the spring of 1971 Mr. Reiss, who was teaching mineralogy that semester, a class I had taken in 1965, decided to offer his students a special field trip, one that fell right in with my interest in pottery making. Mr. Reiss decided to use the pottery glazing process to demonstrate the colors of various minerals when heated.

The commercial glazes I had been using until this time had silica, essentially ground glass, as the medium that was tinted for a particular color. Mr. Reiss' idea was to use borax as the medium, with a ground mineral imparting the color. Essentially we were going to use pots as large borax bead tests. To get ready, Mr. Reiss and I both made a large number of unglazed pots, his designs being more sophisticated than mine since he used a potter's wheel to make his pots.

What we did on a field trip to the Mojave Desert was to have the students find and identify different minerals from several different locations. Once we were at our evening campsite it was my job to dig the firing pit, start the fire, and get the pots ready for the students.

The students would grind up the minerals they had found using a mortar and pestle, and then mix the mineral powder with borax and water. The resulting mixture would be painted on a pot.

After the borax-based glazes had dried, I would put the pots into the fire. The borax would melt and take on the color of the predominant element in the mineral. For instance, uranium minerals will give off a yellow color, iron minerals a red color. Copper minerals such as malachite gave the most spectacular results, yielding a variety of colors depending on the extent of oxidation in the fire, going from green through red to finally metallic copper. The metallic copper resulted when heating a glaze in an oxygen-deprived environment, such as when the pot is surrounded by smoldering coals, called the "raku" process. The students would be in awe of the pots they had painted with the mineral-based glazes.

Student Pouring Borax Glaze on Pot
Spring 1971

Steve with Collection of Finished Pottery

Once the firing of all the pots was completed, a grill would be put over the fire and steaks and hamburgers and hotdogs cooked. It was a very successful and effective and fun teaching tool for basic mineralogy students, and a process that had been used for generations by the Hopi Indians. I still have a collection of pots I made over the years using both the kiln and fire pit heating methods.

At the suggestion of a lady friend, in the summer of 1972, I decided to take my pottery up a notch by making hangers for some of my larger, bowl-shaped pots. The idea was that they could be used for hanging artificial plants. I went to the craft shop and bought a booklet on macramé and a roll of jute. It took a little while to get the making of macramé hangers down, but I was eventually able to make some really nice ones. I made more bowl-shaped pots, all made with coils, and started giving away the pots and hangers to women friends, which they really appreciated.

I quit making pots in the summer of 1972 after graduating from Fullerton Junior College; when I left Fullerton I lost access to a kiln in which to fire pottery. I continued to make macramé hangers until late 1975 when I ran out of ready-made pots suitable for hanging.

After leaving Fullerton I spent a year at Long Beach State University, than moved the San Diego in the fall of 1973 to finish my education. I didn't have much time for hobbies after that; I was working on my Masters Degree in Geology and working full time as a geologist at the County of San Diego, so my time for hobbies was limited. One of the last pots and hangers I gave away in the summer of 1975 went to a woman name Kathy Solis, who two years later became my wife.

Part of Steve's Pottery Collection

EPILOGUE

By the middle of 2000 I was getting old, too old to continue doing some of the work I had been doing at the Kerr-McGee Cushing remediation site. I was already in my middle fifties, older than the CEO and other Kerr-McGee management. I became aware of the fact that, at my age, some of the remediation work I was doing was having a harmful effect on my health as recognized by health and safety personnel. Wearing coveralls impervious to contamination and a respirator causes your body to get easily overheated. This was causing my blood pressure and pulse rate to rise to dangerous levels.

By early 2001 Kathy and I started talking seriously about my early retirement from Kerr-McGee. Kathy was sick, and by then we already knew her prognosis: she was in treatment for cancer and getting noticeably sicker. She encouraged me to retire so that I could spend more time with her.

By May of 2001 we already knew that Kathy was dying; how long she had was the only unknown. In December 2000 she was told "maybe a year." By this time, knowing that I was losing Kathy, it was time to quit my job and spend time with her and care for her.

I filed for early retirement in late May 2001, and quit working on June 7. My retirement included a clause that I could work for Kerr-McGee as a contractor, something many valued employees have done after they retired.

Kathy and I had always looked forward to my retirement when we could finally spend more time together, but it was too late. Kathy died two weeks after I retired. Now I regret all the time spent away from her on projects – I can't go back and live my life with Kathy over again.

STEVEN LOWER'S RESUME

Twenty-seven years of experience demonstrating a high level of proficiency in hydrogeology, project management, environmental planning and compliance analysis, and the environmentally responsible development and protection of natural resources in both private industry and government.

Military Service
United States Army 1966 - 1968

Heavy- and Light-Weapons Infantry and Demolitions, 11th Infantry Brigade and 499th Transportation Battalion
Service in Southeast Asia; Honorably Discharged as Sergeant E-5

Education

- Master of Science, San Diego State University, Geology (Emphasis in Hydrogeology), 1977
- Bachelor of Science, San Diego State University, Geology (Emphasis in Hydrogeology), 1974
- Applied Arts, Fullerton Community College, Science (Emphasis in Geology & Environmental Science), 1972

Professional Experience as an Environmental Management Specialist in Hydrogeology with the County of San Diego, California
1974 - 1979

Responsible for performing geologic and hydrogeologic studies of Subregional and Community Planning Areas as part of a multi-disciplinary team implementing the California Environmental Quality Act (CEQA) in San Diego County. Responsible for land use management, planning, and project analysis as related to geologic, hydrogeologic and other environmental limitations. Duties included, but were not limited to:

- Preparation of maps identifying geology, soils, geologic hazards and natural hazards;
- Analysis of climate, drainage basin runoff, groundwater recharge and storage characteristics, watershed population holding capacity, and groundwater aquifer testing;

- Preparation of recommendations and procedures for the protection of natural resources;
- Manager of multi-disciplinary team responsible for review of County- and privately-initiated Environmental Impact Reports in compliance with the California Environmental Quality Act.
- Preparation of the Groundwater Policy portion of the San Diego County Regional Growth Management Plan providing for the management of groundwater resources;
- Preparation of implementing policies for the San Diego County Regional Growth Management Plan establishing development density controls, a uniform criteria for groundwater resource evaluation and geologic investigations, and the formulation of land use policies based upon watershed population holding capacity;
- Preparation of guidelines setting minimum standards for the investigation of groundwater availability in Subregional and Community Planning Areas;
- Investigation of all projects proposed in areas not served by a municipal water district for compliance with interim long-term groundwater availability criteria;
- Review of all groundwater hydrology reports processed through San Diego County for accuracy and compliance with interim groundwater evaluation guidelines;
- Assisted the public in the identification and development of groundwater resources and the design and maintenance of individual water-supply wells and water-storage systems;
- Represented the County of San Diego at public meetings and made presentations on groundwater availability and geotechnical issues to the public;
- Presented the results of hydrogeologic investigations to, and represented the Environmental Analysis Division and Regional Growth Management at, the San Diego County Environmental Review Board, Planning Commission, and Board of Supervisors.

Professional Experience as a Hydrogeologist with the
Kerr-McGee Corporation
Oklahoma City, Oklahoma
1979 – 2001 (Retired July 2001)

Responsible for all geologic and hydrogeologic activities as related to Kerr-McGee's Wyoming surface coal mine operations and for investigations, assessments and recovery of contaminated and hazardous materials from sites such as oil refineries, petroleum product terminals, natural gas and oil production fields, nuclear materials processing facilities, fertilizer plant sites, and gasoline service stations. Primary assignment as project hydrogeologist at the Cushing, Oklahoma oil refinery remediation site 1999 to 2001. Duties included, but were not limited to:

- Preparation of coal mine hydrogeology reports, special reports addressing specific issues regarding the hydrogeology of aquifers affected by surface coal mining and support activities, and portions of coal mine permit documents involving hydrogeology;
- Geologic mapping and interpretation;
- Analysis of climate, drainage basin runoff, groundwater recharge and storage characteristics, and groundwater aquifer testing;
- Environmentally responsible development of groundwater resources to support company projects;
- Design and construction of high-capacity facility water-supply wells for operational, environmental management of remediation sites, irrigation, and fire control purposes;
- Design, construction and maintenance of domestic (homeowner) water-supply wells installed to replace wells impacted by contamination;
- Plugging and abandonment of water-supply wells, monitor wells, hazardous waste injection wells and oil wells;
- Design, implementation and supervision of hydrogeologic site investigations;
- Design, installation and management of groundwater monitoring programs;
- Design and installation of oil, gasoline and acidic groundwater interception and recovery systems at oil refinery sites and gasoline stations;
- Design and installation of groundwater interception and dewatering systems;
- Performance and analysis of single- and multiple-well short-term and long-term aquifer tests, field well-bore tests, water-quality tests, and subsurface geophysical investigations;
- Design, management and field direction of land-based and barge-based contaminated oil refinery soil and waste characterization investigations;
- Design and performance of hydrogeologic field reconnaissance investigations for waste disposal cells;
- Computer modeling of groundwater regimes;
- Critical review of technical reports prepared by in-house staff and outside contractors;
- Often assigned as "trouble-shooter" to resolve management and technical problems with other Kerr-McGee Corporation staff member's and outside contractor's projects;

- Supervision of Kerr-McGee Corporation staff members, hiring and management of contractors, and the accounting of funds necessary to complete projects such as the assessment of hazardous waste sites, construction of monitoring and water-supply wells, groundwater monitoring programs, treated waste disposal cell siting investigations, plugging and abandonment of oil wells and hazardous waste injection wells, and oil refinery waste characterization projects;

Professional Experience as a Teacher

- Responsible for the preparation and instruction of General Geology Laboratory classes in the Graduate Student Instructor Program at the San Diego State University Department of Geology, 1975 – 1977;
- Responsible for the preparation and instruction of General Geology, General Geology Laboratory, Historical Geology, and Field Geology at the San Diego Mesa Community College Department of Geology, 1977 – 1979. Holds a Life-Time California Community College Teaching Credential;
- Volunteer lecturer on the physical sciences at elementary and middle schools, 1975 – 1979;
- Volunteer Judge at Middle School Science Fairs, 1975 – 1981.

www.ingramcontent.com/pod-product-compliance
Lightning Source LLC
Chambersburg PA
CBHW070247190526
45169CB00001B/331